AutoCAD 2016 from Zero to Hero

AutoCAD 2016 from Zero to Hero

Ali Akbar
Zico Pratama Putra

Kanzul Ilmi Press
2017

First Printing: 2017

ISBN-13: 978-1546346494

ISBN-10: 154634649X

Kanzul Ilmi Press
Woodside Ave.
London, UK

Bookstores and wholesalers: Please contact Kanzul Ilmi Press email

zico.pratama@gmail.com.

Trademark Acknowledgments

Ordering Information: Special discounts are available on quantity purchases by corporations, associations, educators, and others. For details, contact the publisher at the above-listed address.

Contents

CHAPTER 1 INTRODUCTION TO AUTOCAD

Welcome to the AutoCAD's World. In this first chapter, I'll introduce you to commands and AutoCAD's user interfaces. But first, you have to know why CAD software is now replaces traditional pencil drawing and now you don't have to use that big, drafting table to draw an advanced drawing.

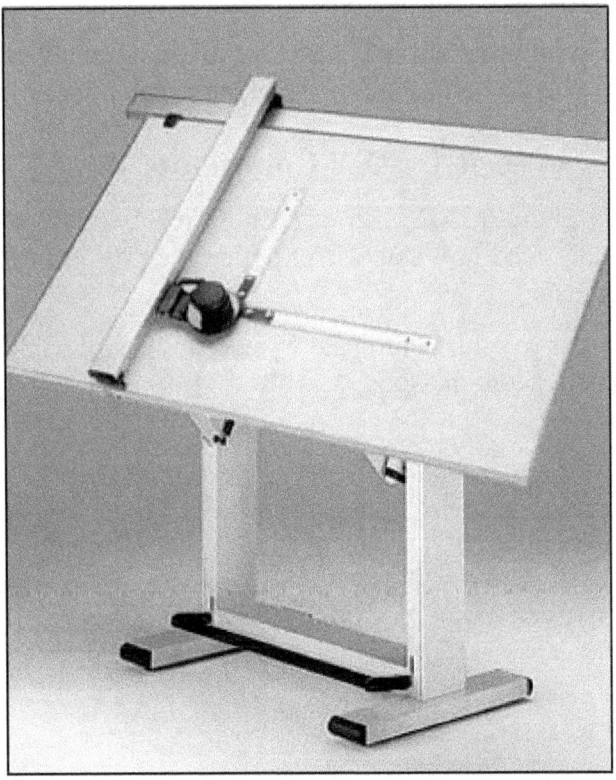

Pic 1.1 Most of current engineering drafter do not have this "ancient" drafting table

These are features of CAD's software that make drawing with software is better:

- Precision: You can draw a line, the arc, and other forms with incredible precission. Accuracy in AutoCAD is 14 decimal point.

- Modifiable: An advanced drawing created a long time ago can be modified again to draw a new drawing. While old drawing in pencil/pen cannot be updated and you have to create the new drawing from scratch.

- Clean: You don't have to own eraser to draw a drawing.

- Efficiency: You can create more drawing in the same time, and you can create drawing faster. Especially when you need repetition, like drawing a multi-storey building, or floor tiles.

- Popular: Everyone uses it.

- Easy to Publish: Because it's digital, you can give the drawing on people across the globe just in a second.

1.1   XY Coordinate.

All objects in AutoCAD is in exact position. Because of that, you have to understand how AutoCAD define the position using simple X, Y coordinate.

AutoCAD has World Coordinate System (WCS). For 3d drawing, there will be Z additional axis.

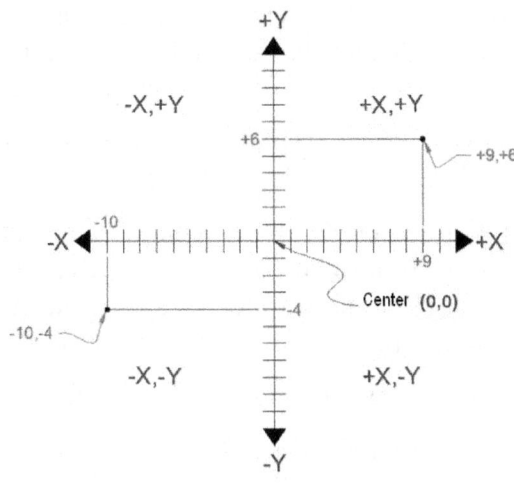

Pic 1.2a Simple XY WCS coordinate used in AutoCAD

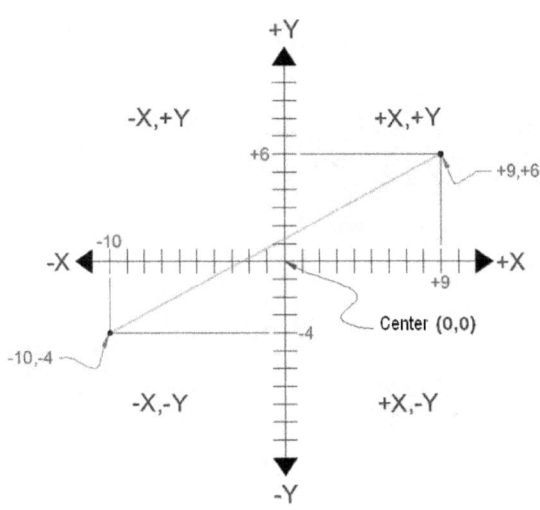

Pic 1.2b A line from -10,4 to e +9, +6

AutoCAD has x,y point to define where the point object located. There is also origin or center point (0,0) where all object positions use this point as an initial reference.

See above pic to see the AutoCAD x, y coordinate and how to draw a line between two coordinates. For example xy coordinate 9,6 means x = 9 and y = 6.

Coordinate (-10,-4), means x = 10 units negative (left side) and y = 4 units negative (below)

Sometimes, you don't know the initial point exact location, you just know that you want to draw to next point relative to this position. You can use relative coordinate by adding symbol @ (SHIFT + 2) to tell AutoCAD that next point is relative to the last point.

Here are some important points about X, Y coordinate.

- The absolute point is the exact position of a point, relative to 0,0.

- The relative point is relative to the last point.

1.2   Angle in AutoCAD.

AutoCAD also has angle to draw. Here's how to specify the angle in AutoCAD:

✓ The X positive is the 0 degree.

✓ Counter clock wise is positive

✓ Clockwise is negative.

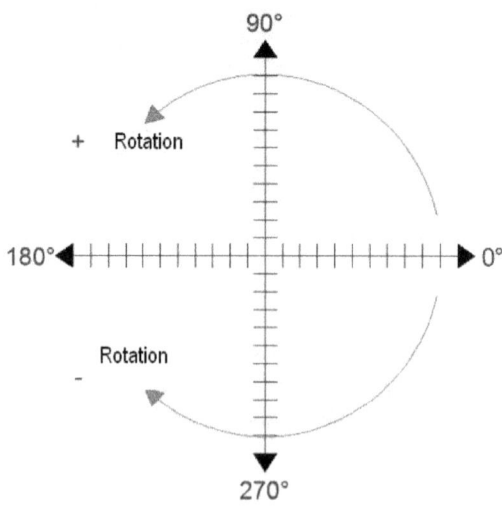

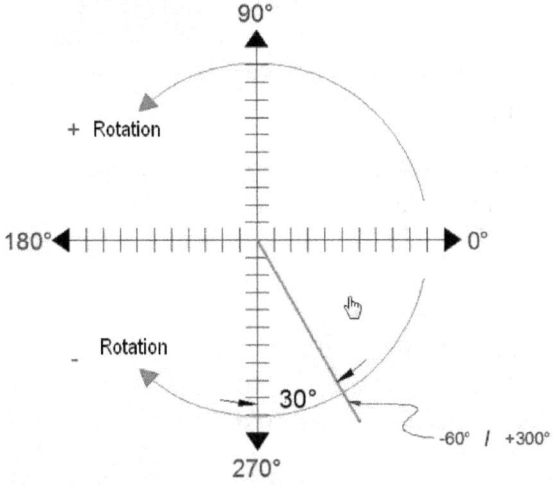

Pic 1.4 Line example created using angle

For example 90 degrees = Y positive.

You can measure angle based on other angle.

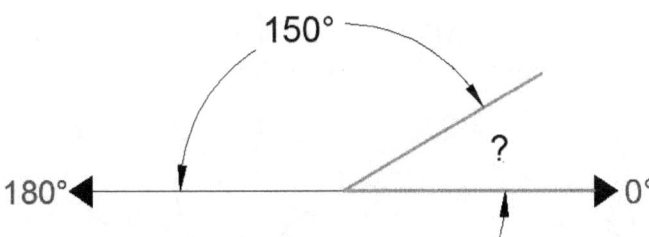

Pic 1.5 Calculating an angle from another angle

From above examples, some important notes:

- 0 degrees on hour = 3 position.
- 180 degrees on hour = 9 position

1.3   Inserting Point in AutoCAD

Here are three methods to insert point in AutoCAD:

- Absolute coordinate: Just insert the xy point relative to the center point (0,0). Insert x value first, and then y value.

- Relative coordinate: Insert by adding prefix @ so you enter @X,Y. This will put point x,y points relative to the last position.

- Polar coordinate: Insert by using template @D<A. Which D is the length and A is the angle: For example @10<90 will draw a line with length = 10 units and 90 degrees direction.

Notes:

- The three methods are the only methods for inserting point in AutoCAD, there are no other methods to draw AutoCAD. X value has to be inserted first, then Y value.

- Don't forget the '@' symbol when you insert relative value. All mistakes in inserting input will generate unintended results.

- If you want to do checking, click F2, and then click F2 again

1.4 AutoCAD's User Interfaces

When you run AutoCAD program for the first time, you can see AutoCAD's window like the picture below:

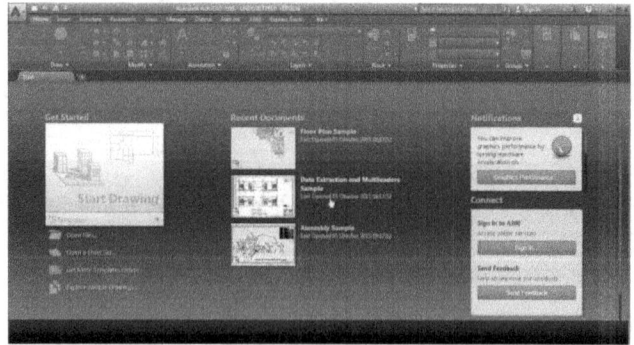

Pic 1.6 Start window

To draw a new file, click Start Drawing, an AutoCAD's user interface for the drawing will be displayed.

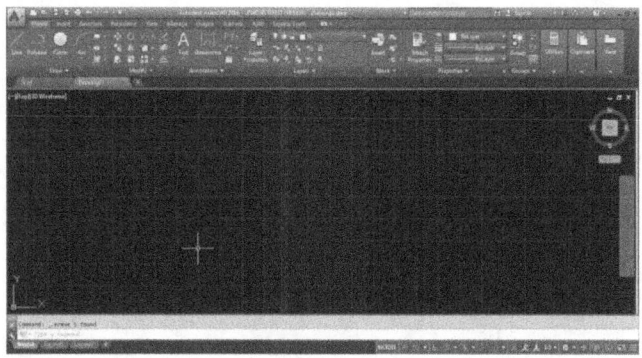

Pic 1.7 Drawing interface

1.4.1 Ribbon

When you on drawing page, buttons in the ribbon will be enabled. The ribbon interface is similar with MS Office interface so you will be familiar and makes drawing process easier.

In HOME tab, you'll see buttons to Draw, and Modify drawing.

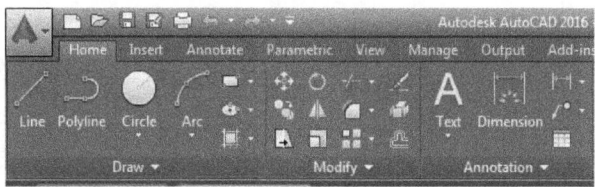

Pic 1.8 Draw and Modify

Still in Home tab, there are Annotation and Layers, the Annotation box used to give annotation to your drawing, eg: text, dimension, etc. The Layers box is to insert layers to your drawing. The user can add a layer to overlay the drawing.

Pic 1.9 Annotation and Layers boxes

There are **Block**, **Properties**, and **Groups** boxes. **Block** contains buttons to block more than one object to become a single object. The

Properties box used to manage the properties of an object. **Groups** to group or ungroup objects.

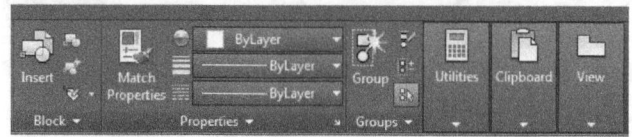

Pic 1.10 Block, Properties and Groups

Insert box is used to insert many kind of objects, from Block, Definition, Reference, Point Cloud and Import.

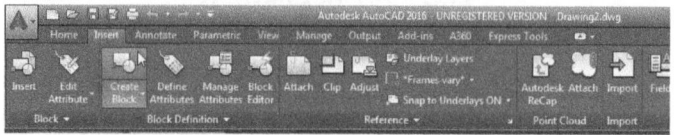

Pic 1.11 Insert tab

Annotate tab used to insert more detail anotate, from texts, dimensions, leaders, etc.

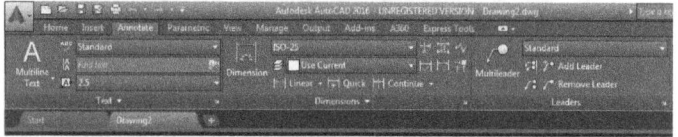

Pic 1.12 Tab Annotate

The parametric tab has buttons that used for managing geometric and dimensional drawing.

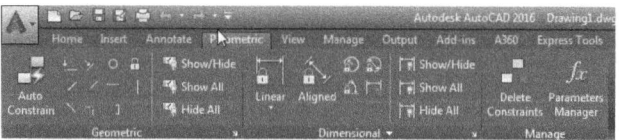

Pic 1.13 Parametric tab

View tab is used to modify the user interface of AutoCAD. You can manage the Viewport, Palettes, and Interface.

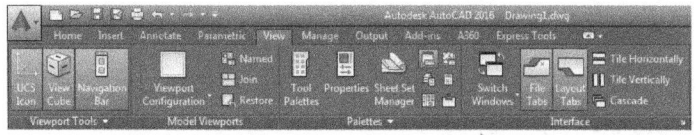

Pic 1.14 View tab from ribbon

Manage tab, used to create macro used to record your action. You can do coding in macro.

Pic 1.15 Manage tab

Output tab used for exporting and printing your drawing to paper or other forms.

Pic 1.16 Tab Output

In Add Ons tab, you can manage add-on applications.

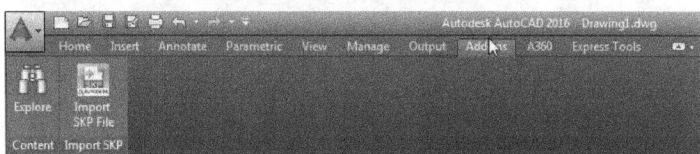

Pic 1.17 Add On application

A360 tab consists of buttons that enable you to use online features of AutoCAD.

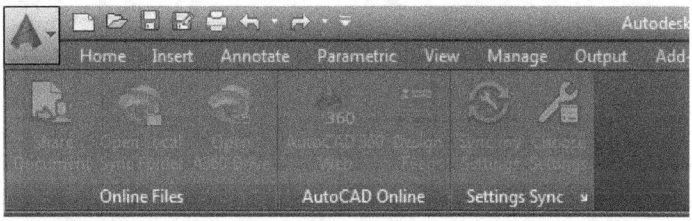

Pic 1.18 Cloud saving function

Express Tools tab can be used to manage blocks, texts, modifying objects and layout.

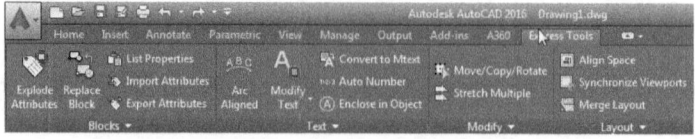

Pic 1.19 Tab Express Tools

Ribbon in AutoCAD can be minimized, just click 2x on the ribbon tab.

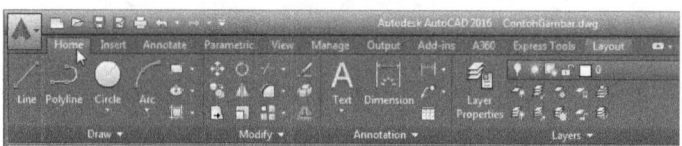

Pic 1.20 Double click on ribbon tab

The ribbon will be minimized.

Pic 1.21 Buttons in ribbon minimized

If you click twice again, the buttons are hidden, and the ribbon only displays the texts.

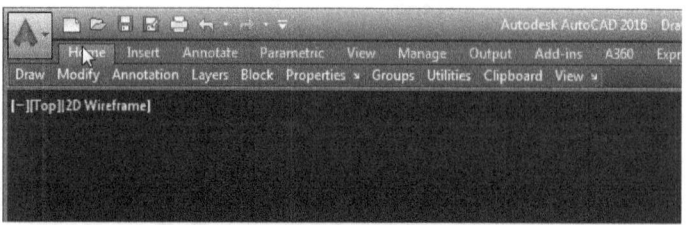

Pic 1.22 Ribbon's buttons hidden

1.4.2 Menus

Main Menus can be opened by clicking A button on the top left of
your AutoCAD window:

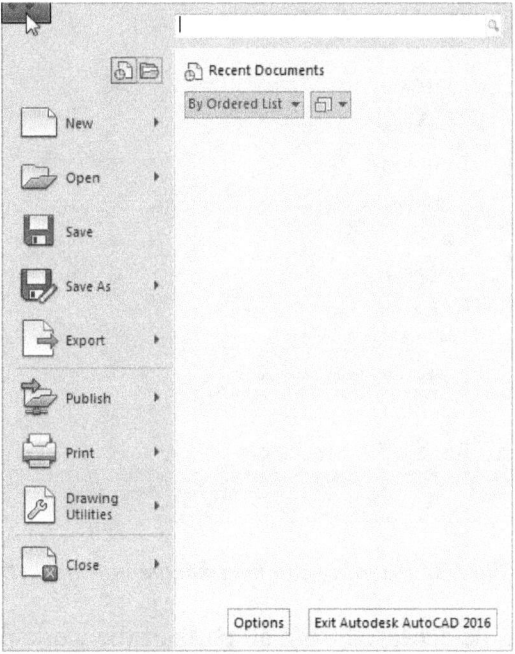

Pic 1.23 Main menu of AutoCAD

On the menu box above, you can see Sethe arch Command textbox
that makes finding commands easier. Just enter the name of the
command, and AutoCAD will autocomplete it for you.

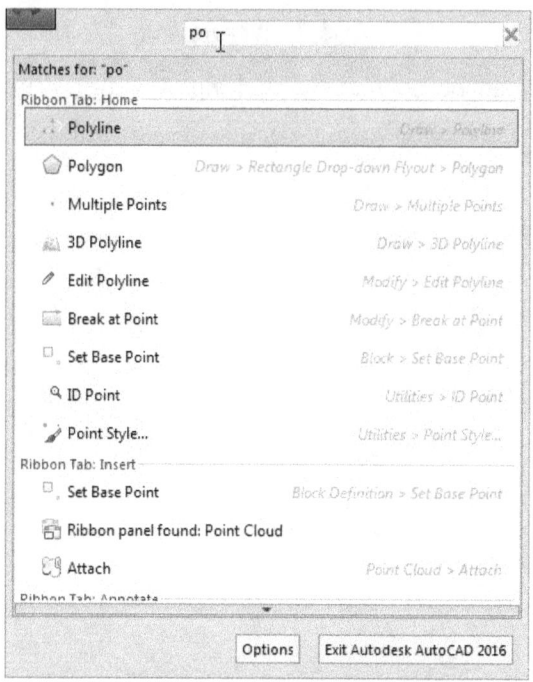

Pic 1.24 Inserting command name in AutoCAD

This menu can be accessed from all parts of the workspace. Menus in the main menu are:

1. New: To draw a new drawing, from template, or create sheet set that manages drawing layouts, paths, and project data.

Pic 1.25 New menu

2. Open : To open existing drawing.

Pic 1.26 Menu Open

3. Save: Save existing drawing changes, if the drawing hasn't saved before, it will save to a new file.

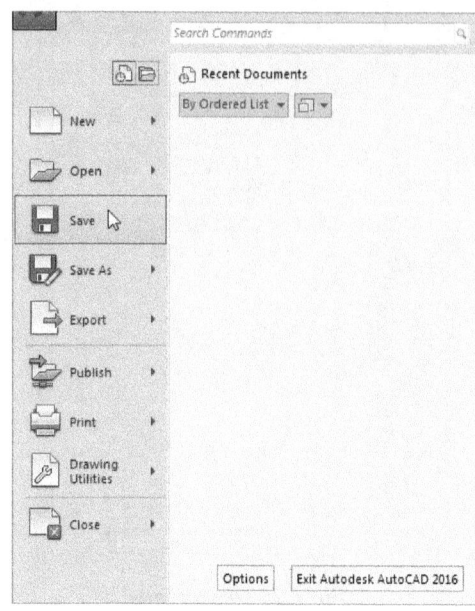

Pic 1.27 Save menu

4. Save As: Save existing drawing to a new file.

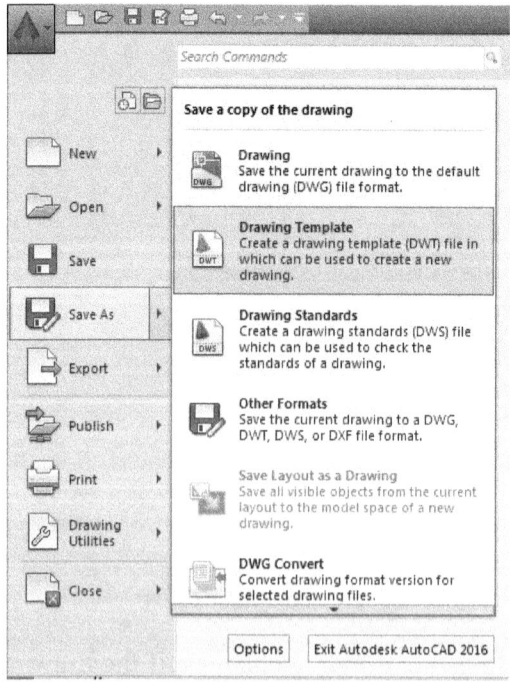

5. Export: Save drawing to other file formats, such as Design Web Format (DWF), PDF, and other CAD files.

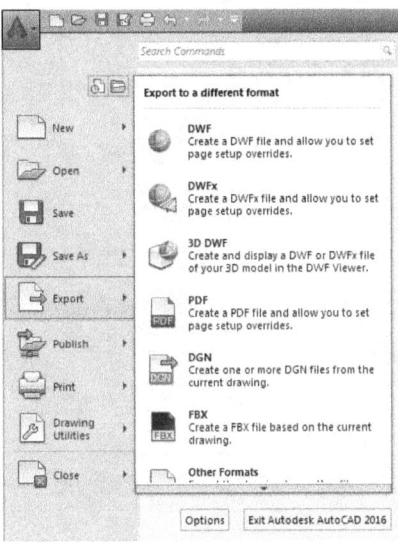

Pic 1.29 Export Menu

6. Publish: Send 3D model to 3D printing service, or create the archived sheet set (AutoCAD LT doesn't support 3D.) etc.

Pic 1.30 Publish menu

7. Print: Printing single drawing, or batch-plot. You can also setup the page and style plot.

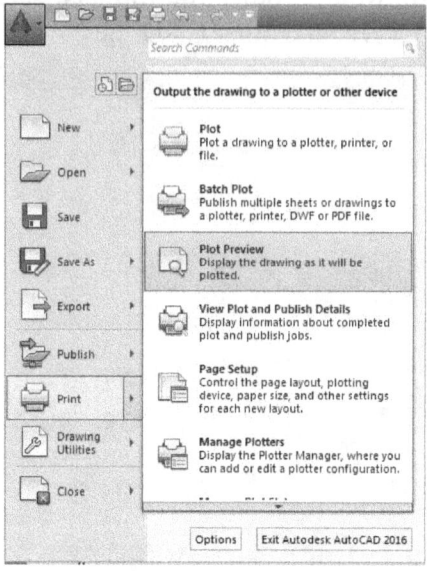

Pic 1.31 Print menu

8. Drawing Utilities: Setting file properties, or drawing unit, doing purging on unused blocks, doing auditing or recovering damaged drawing.

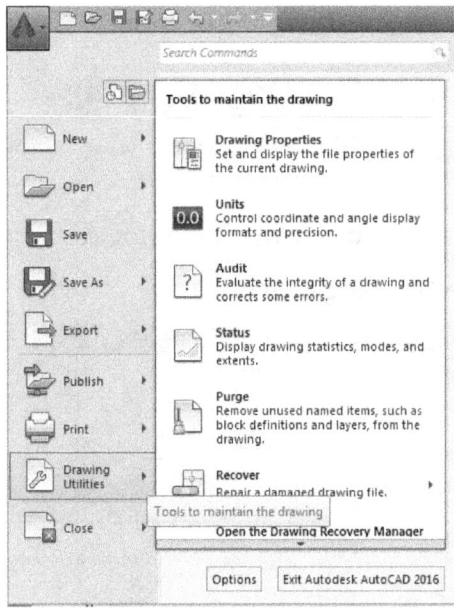

Pic 1.32 Drawing Utilities

9. Close: Closing existing drawing, if the drawing already modified and hasn't been saved, this will generate Save confirmation box.

1.5   Open Drawing

You can open drawing file to display on your AutoCAD using steps below:

1. Click on AutoCAD icon to display AutoCAD:

2. Click **Open > Drawing** menu.

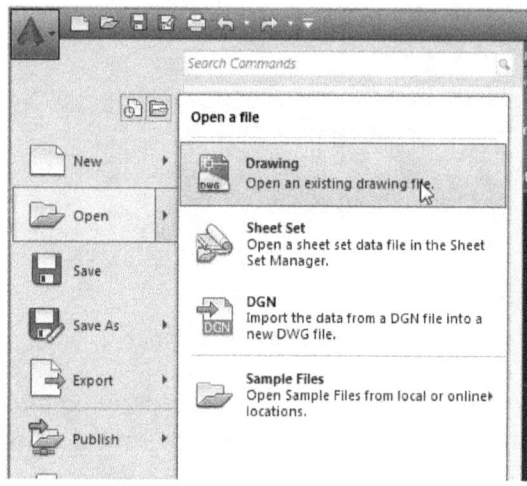

Pic 1.34 Click Open > Drawing

3. **Select File** window appears, choose the file you want to open, click **Open**.

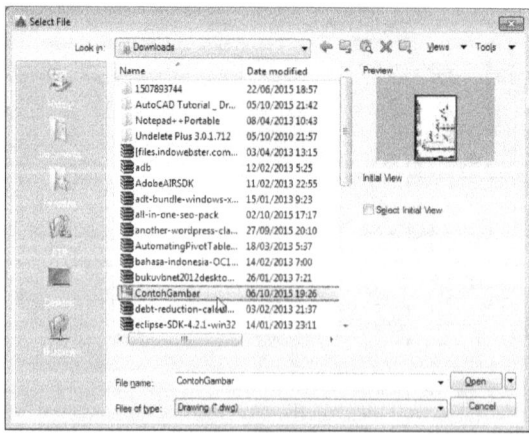

4. The drawing will be opened.

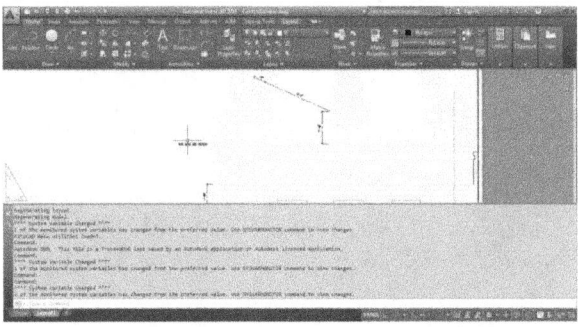

1.36 Drawing opened in AutoCAD

5. AutoCAD can display more than one drawing. Each drawing will be opened as MDI (multiple document interface) windows.

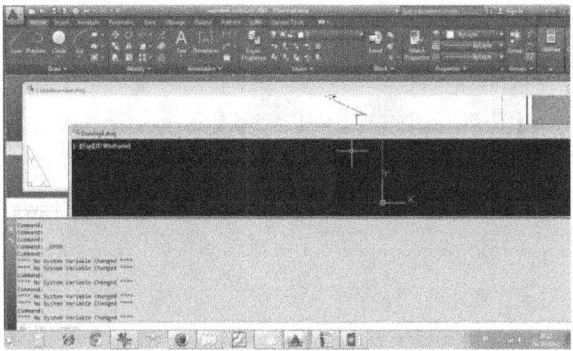

Pic 1.37 AutoCAD may open more than one project

1.6   Close Drawing

A drawing that doesn't need to be edited further, close it using steps below:

1. Click AutoCAD icon to open the main menu.

2. Click **Close > Current Drawing** menu to close the active drawing.

3. Or click **Close > All Drawing** to close all drawing.

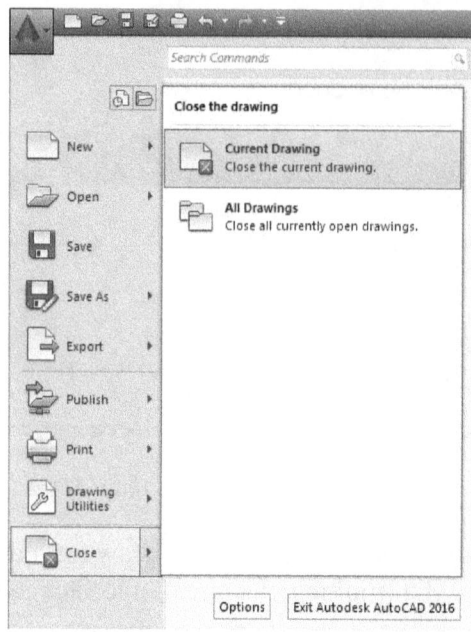

Pic 1.38 Close menu to close the drawing

4. If your modification hasn't yet saved, a confirmation window will appear and ask whether you want to save it or not. Click Yes to save and No if you don't want to save the modification.

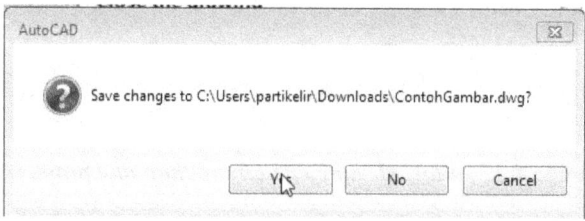

Pic 1.39 Confirmation window

1.7   Export as PDF

PDF (portable document format) is the most pervasive format used in the design world. AutoCAD can export it's drawing directly to pdf without the third-party software or add-in.

Look at steps below to export your drawing as PDF:

1. Click on AutoCAD icon.

2. Click on **Export > PDF** menu.

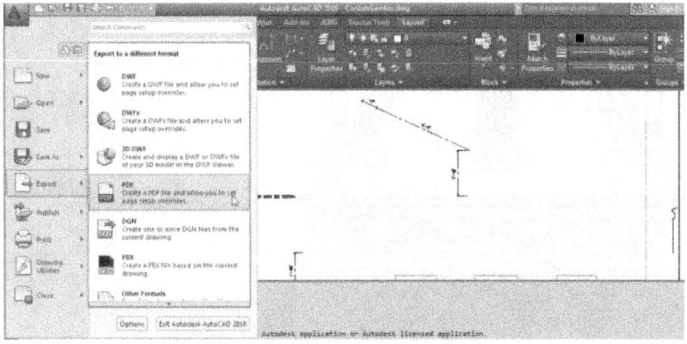

Pic 1.40 Menu Export > PDF

3. **Save As PDF** window emerges, choose a filename for the new pdf file in **File name** textbox. And click **Save**.

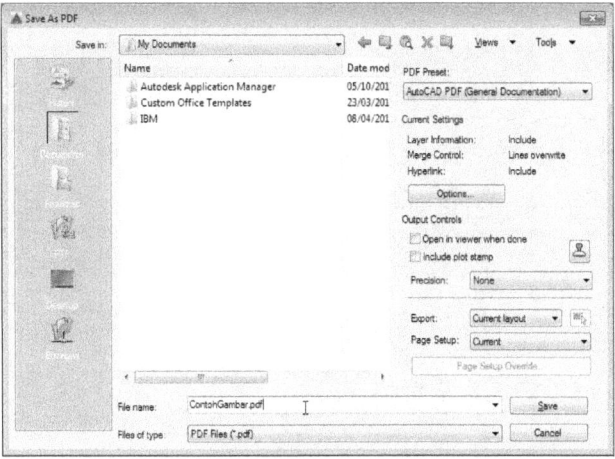

Pic 1.41 Inserting file name for pdf file

4. The file will be inserted as PDF.

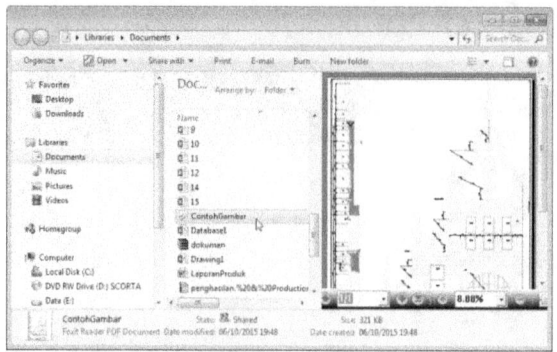

Pic 1.42 PDF file already created

CHAPTER 2 DRAWING IN 2D

In this chapter, I'll explain important commands that let you draw 2 dimensional drawing in AutoCAD. Drawing in 2D is the basic of AutoCAD drawing.

2.1 Create 2D Drawing

There are lots of 2d drawing types you should understand. From Line to Donut. I'll show you how to draw 2d using those types of drawing.

2.1.1 Draw a line

Line is the basic type of drawing. It's a straight line that connects two dots. You can draw a line by following steps below:

1. Click **Line** button in **Home > Draw** ribbon.

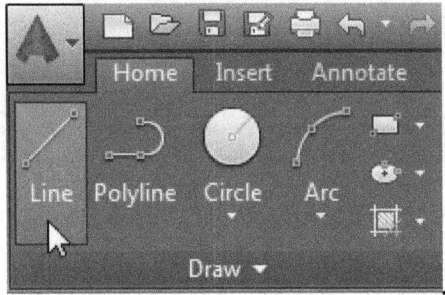

Pic 2.1 Click Line button in Home > Draw

2. Or you may type "line" in the command prompt.

3. Command prompt appears:

```
LINE from point:
```
4. You can insert with absolute coordinate or click on the drawing.

```
Prompt: To point:
```
5. Insert the second point location.

6. Before creating LINE, limits your workspace by inserting LIMITS command.
```
Command: LIMITS
Reset Model space limits:
```
7. Set the bottom-left limit to 0,0.
```
Specify lower left corner or [ON/OFF] <0.0000,0.0000>: 0,0
```
8. Then specify the top right limit to 100,100. This will make creating picture easier, because the canvas for this tutorial is from 0,0 to 100,100.
```
Specify upper right corner <420.0000,297.0000>: 100,100
```
9. Then type the line to start creating line, specify the first point to 10,10.
```
Command: LINE
Specify first point: 10,10
```
10. When you move the pointer, you'll see that the first point of line glued to 10,10, and you still can move the mouse pointer.

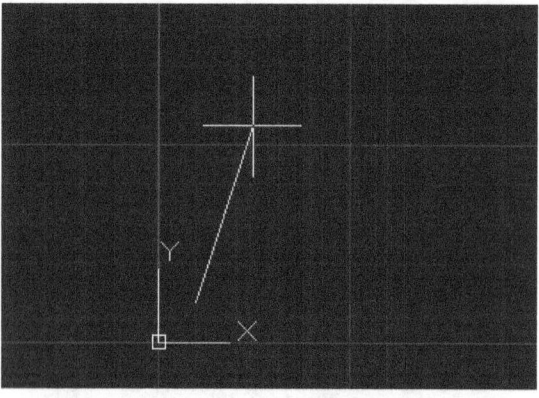

Pic 2.2 The first point of line glued to 10,10

11. You can change the pointer right or left.

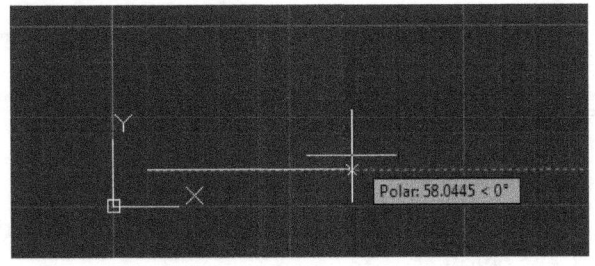

Pic 2.3 Pointer mouse still can be moved

12. For the next point, choose 50,50. You can see the line glued to 50,50.

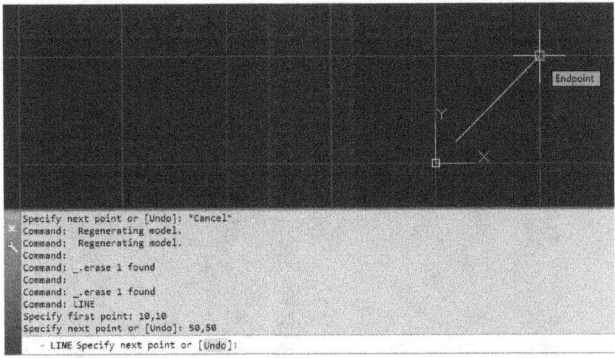

Pic 2.4 Line connected from 10,10 to 50,50

13. Click Enter, a line will be created, and the pointer released from the line.

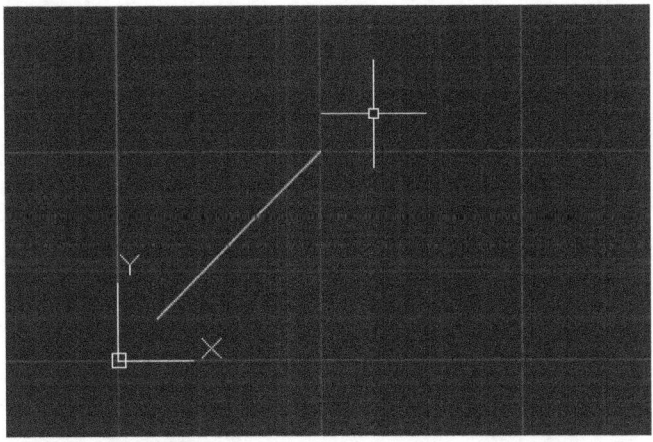

Pic 2.5 Line created, and the pointer mouse released

14. All command line texts in this tutorial:

```
Command: LINE
Specify first point: 10,10
Specify next point or [Undo]: 50,50
Specify next point or [Undo]:
```

In the next tutorial, we'll draw a line using relative coordinate, you can see the steps below:

1. Type Line

```
Line
```

2. Specify the first point = 10,10.

```
Specify first point: 10,10
```

3. Specify next point @50,25 relative from the first point.

```
Specify next point or [Undo]: @50,25
```

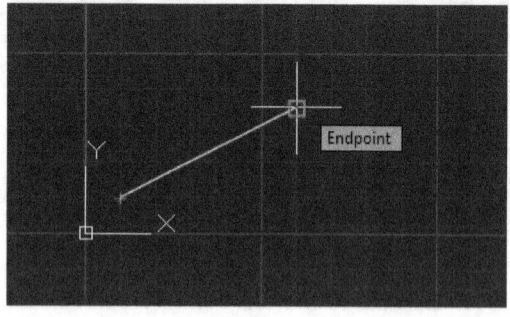

Pic 2.6 Specify second point using relative coordinate

4. Click Enter, line will be created.

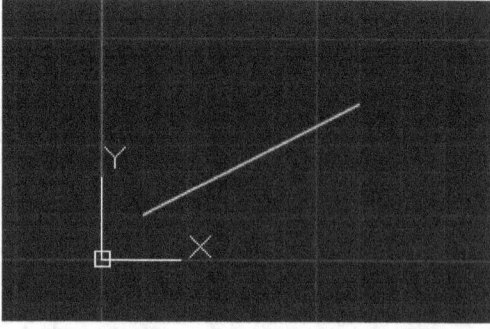

Pic 2.7 Line created using relative coordinate for the second point

5. To delete line, click on the line to select the line first. Selected line will become dotted line.

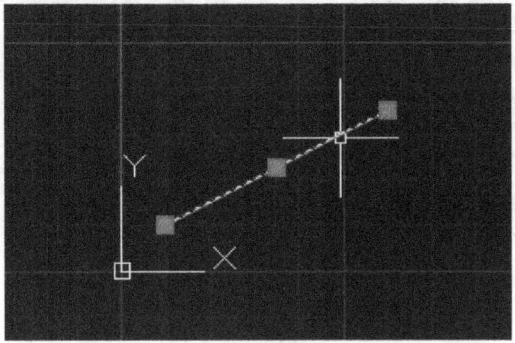

Pic 2.8 Selected Line become dotted

6. Click **Delete** button on your keyboard, or right-click and click **Erase** menu.

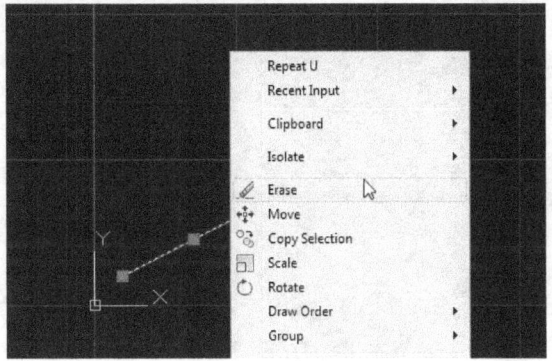

Pic 2.9 Erase menu to delete selected line

The third tutorial is using angle coordinate. Here's the method to draw a line using angle coordinate:

1. Insert line command, and specify the first point = 10,10.

```
Command: LINE
Specify first point: 10,10
```

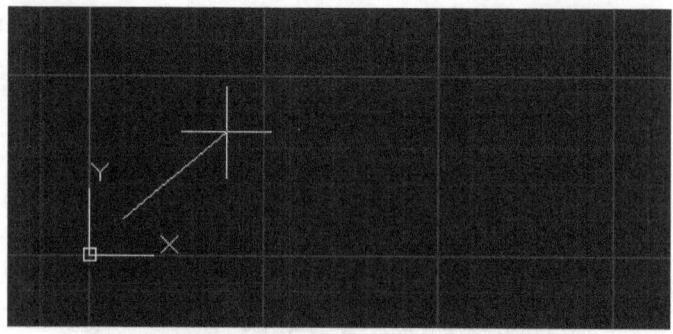

Pic 2.10 Specify the first point = 10,10

2. Then Specify second point 50 units from the first point, and with <45 degrees. Click Enter:

```
Specify next point or [Undo]: @50<45
Specify next point or [Undo]:
```

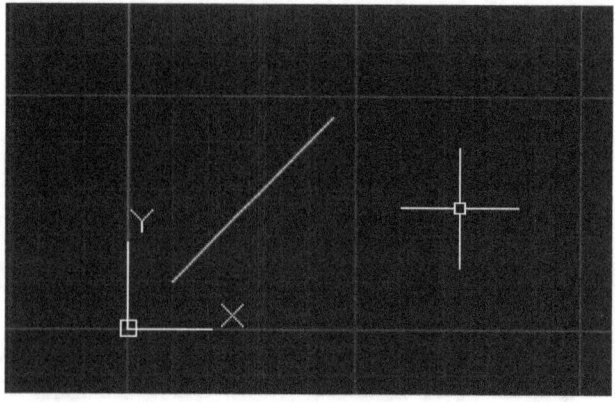

Pic 2.11 Drawing line by degree coordinate

2.1.2 Draw Polyline

Polyline is multiline, more than one lines that composed by line and the arc segments. See picture below for polyline example:

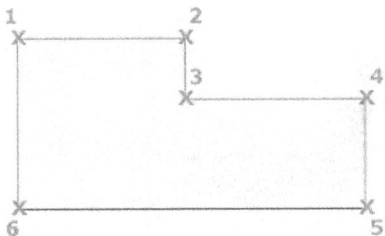

Pic 2.12 Polyline example

Some notes on polyline:

- Specify start point: similar to LINE command, specify the first point or initial point.

- Next:

```
polyline, line, or the arc.
Specify next point
```

- If you choose the second point, you'll create straight line.

- If you enter other option, for example the arc, you'll make an the arc.

There are some prompts related to line and the arc:

- Close : connects first segment and the last segment to draw a closed polyline.

- Halfwidth: half width of the segment, from the center to outer.

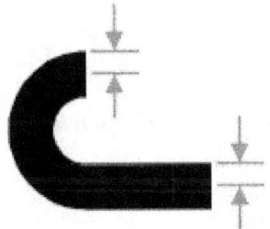

Pic 2.13 Halfwidth

- Width: The width of next segment.

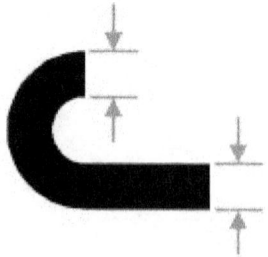

Pic 2.14 Width

- The first part of width will equal the last width. The last width will be uniform to all segments until you change another width. The first part and the end part of width is similar to the width at the middle of the segment.

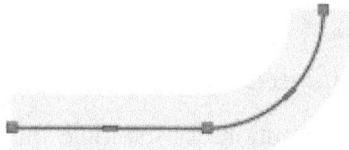

Pic 2.15 Width in line and the arc

- At the intersection of a polyline, there will be a bevel.

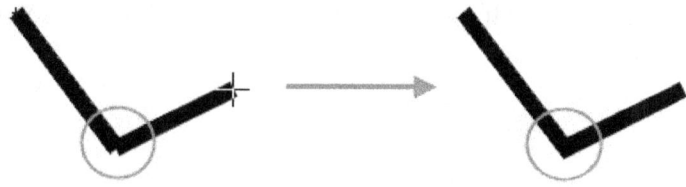

Pic 2.16 Beveling in polyline

- Undo will erase the last segment added.

Some arguments in Line-Only prompt:

- The arc: creating the arctangent from the previous segment.

- Length: Creating segment with length = next segment. If the next segment is an the arc, the new segment will be tangent from the arc segment.

Pic 2.17 Length

Some arguments in the arc-only prompts:

- Endpoint of the arc: Completing the arc segment. Tangent from the previous polyline segment.

- Angle: Specifying the angle from the arc segment from center point. If positive = counter clockwise, if negative = clockwise.

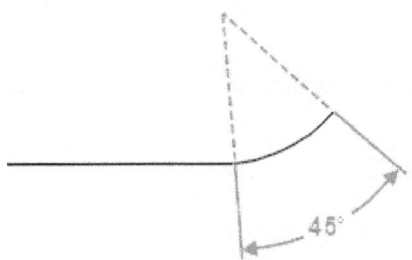

Pic 2.18 Angle

- Center: Specifying the arc segment based on center point. See picture below:

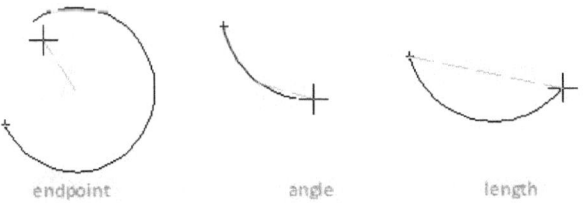

endpoint   angle   length

- Direction: Specifying tangent for the arc segment.

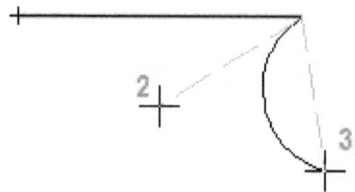

Pic 2.20 Direction

- (2) is the tangent direction from the arc's start point.
- (3) is the last point of the arc. You can use Ctrl to draw counter clockwise.
- Line: Change from the arc drawing to line drawing.
- Radius: Determine the radius from the arc segment.
- Second pt: Determine the second point and the last point from three points the arc.

For a Linetype pattern look at the arguments below:

- PLINEGEN system variable, determine what type of line created in 2 dimensional polyline.
- 0 will create dash at the corner.
- 1 will draw an uninterrupted dotted line.

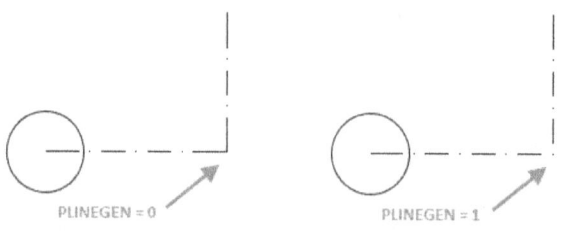

Pic 2.21 The difference between PLINEGEN = 0 and PLINEGEN = 1

See tutorial below to create Polyline:

1. First, create limit from 0,0 to 100,100.

```
Command: limits
Reset Model space limits:
Specify lower left corner or [ON/OFF] <0.0000,0.0000>: 0,0
Specify upper right corner <420.0000,297.0000>: 100,100
```

2. Type polyline and specify the first point to 0,0.

```
Command: POLYLINE
PLINE
Specify start point: 10,10
Current line-width is 0.0000
```

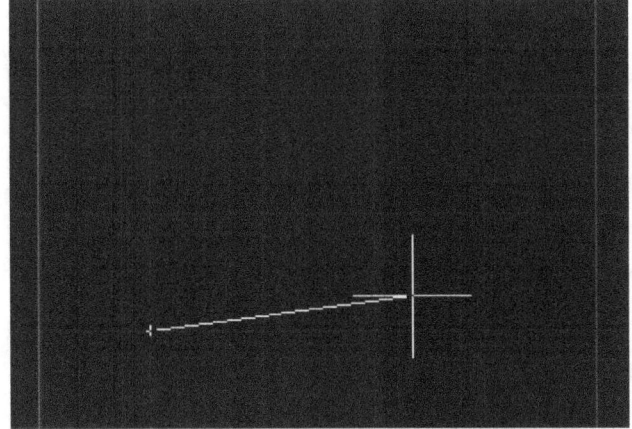

Pic 2.22 Specify the first point of Polyline to 10,10

3. Specify next line to 80,10.

```
Specify next point or [The arc/Halfwidth/Length/Undo/Width]:   <Ob-
ject Snap Tracking on> 80,10
```

Pic 2.23 Specify second point to 80,10

4. Specify next point to 80,50.

```
Specify next point or [The arc/Close/Halfwidth/Length/Undo/Width]:
80,50
```

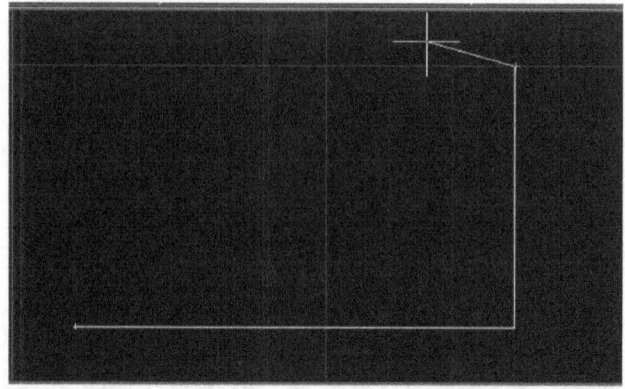

Pic 2.24 Specify next point to,50

5. To create the the arc, insert argument A, and specify angle = 20 and end point of the arc to 10,50.

```
Specify next point or [The arc/Close/Halfwidth/Length/Undo/Width]:
A
Specify endpoint of the arc (hold Ctrl to switch direction) or
[Angle/CEnter/CLose/Direction/Halfwidth/Line/Radius/Second
pt/Undo/Width]: A
Specify included angle: 20
Specify endpoint of the arc (hold Ctrl to switch direction) or
[CEnter/Radius]: 10,50
Specify endpoint of the arc (hold Ctrl to switch direction) or
```

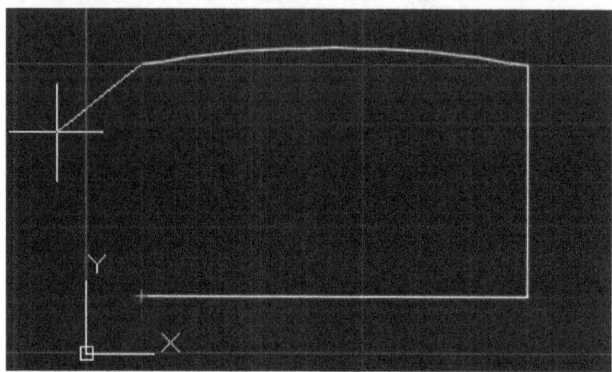

Pic 2.25 The arc creation

6. Back to draw a line, by inserting 1 argument and type C to close the polyline.

```
Specify endpoint of the arc (hold Ctrl to switch direction) or
[Angle/CEnter/CLose/Direction/Halfwidth/Line/Radius/Second
pt/Undo/Width]: l
Specify next point or [The arc/Close/Halfwidth/Length/Undo/Width]:
C
```

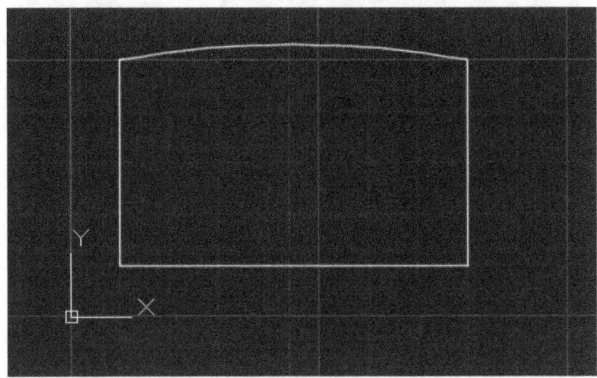

Pic 2.26 Polyline creation

The second tutorial on creating polyline:

1. Insert polyline set start point to 10,10 set halfwidth to 1.

```
Command: POLYLINE PLINE
Specify start point: 10,10
Specify next point or [The arc/Halfwidth/Length/Undo/Width]: h
Specify starting half-width <5.0000>: 1
Specify ending half-width <1.0000>:
```

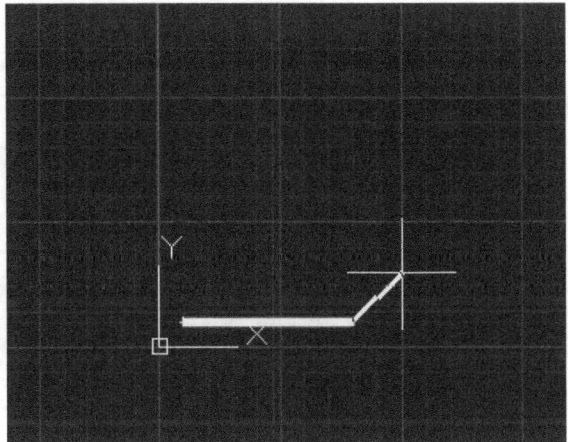

Pic 2.27 Setting halfwidth to 1 and start point to 10,10

2.   Set the next point to 80,10 and 80,50.

```
Specify next point or [The arc/Halfwidth/Length/Undo/Width]: 80,10
Specify next point or [The arc/Close/Halfwidth/Length/Undo/Width]:
80,50
```

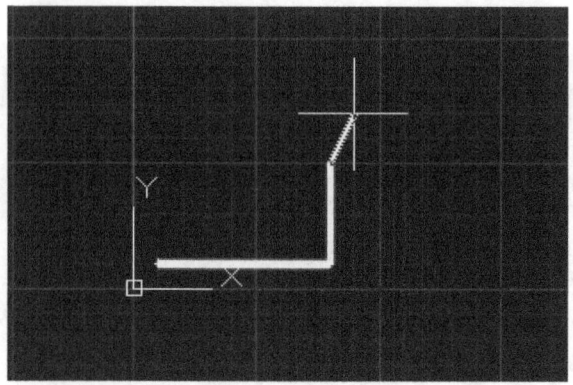

Pic 2.28 Set next point to 80,10 and 80,50

3. Create the arc with radius = 50 and the next point to 10,50.

```
Specify next point or [The arc/Close/Halfwidth/Length/Undo/Width]:
a
Specify endpoint of the arc (hold Ctrl to switch direction) or
[Angle/CEnter/CLose/Direction/Halfwidth/Line/Radius/Second
pt/Undo/Width]: r
Specify radius of the arc: 50
Specify endpoint of the arc (hold Ctrl to switch direction) or [An-
gle]: 10,50
```

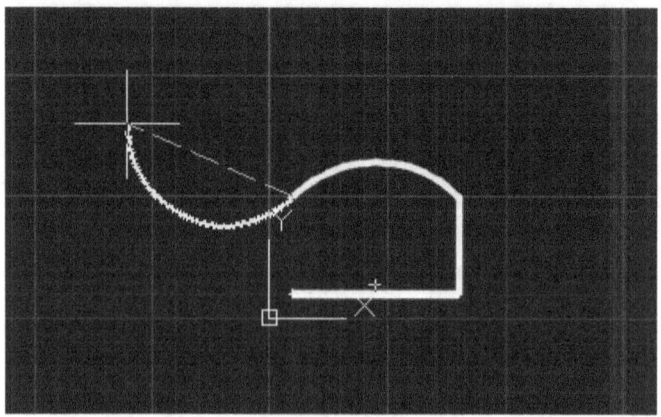

Pic 2.29 Create the arc

4. Choose Close, polyline created with width = 2.

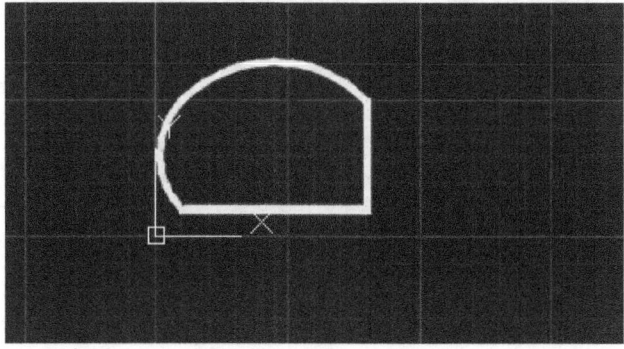

Pic 2.30 Polyline created

2.1.3 Draw a circle

Circle command used to draw a circle, you can make a circle using some combinations. See examples below to draw a circle in AutoCAD:

1. Type circle, command and let the center to 50,50.

```
Command: CIRCLE
Specify center point for circle or [3P/2P/Ttr (tan tan radius)]:
50,50
```

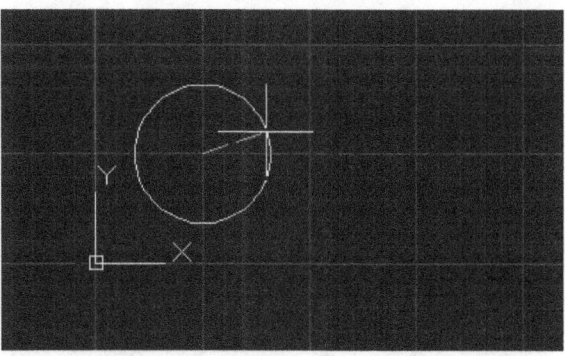

Pic 2.31 Specify center point to 50,50

2. Specify the radius = 50. A circle will be created

```
Specify radius of circle or [Diameter] <50.0000>: 50
```

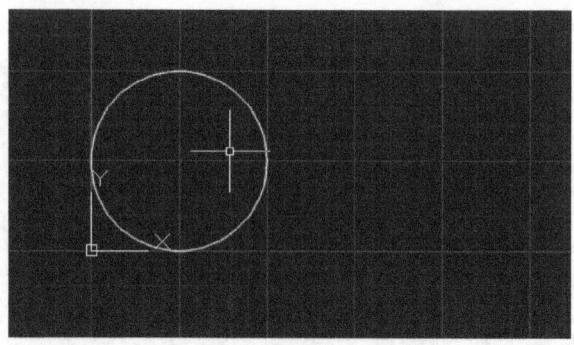

Pic 2.32 Draw a circle with center = 50,50 and radius = 50

You can also draw a circle by specifying three points. Look at this tutorial:

1. Insert circle command and chooes 3p.

2. Specify the first point to 50,0, the second point to 100,0 and the third point to 50,50.

```
Command: CIRCLE
Specify center point for circle or [3P/2P/Ttr (tan tan radius)]: 3p
Specify the first point on circle: 50,0
Specify second point on circle: 100,0
Specify the third point on circle: 50,50
```

3. AutoCAD will draw a circle based on three points inserted.

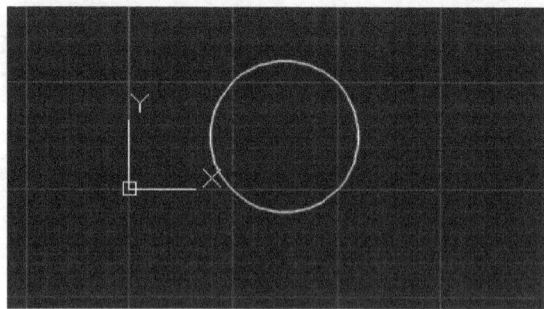

Pic 2.33 Circle created by specifying 3 points

You can also specify 2 points to draw a circle. See steps below:

1. Type "circle" and type 2P for 2 points.

2. Specify the first point 10,10 and 100,10 as second point.

```
Command: CIRCLE
Specify center point for circle or [3P/2P/Ttr (tan tan radius)]: 2p
Specify first end point of circle's diameter: 10,10
Specify second end point of circle's diameter: 100,10
```

3. If it's saved, you'll see the circle created:

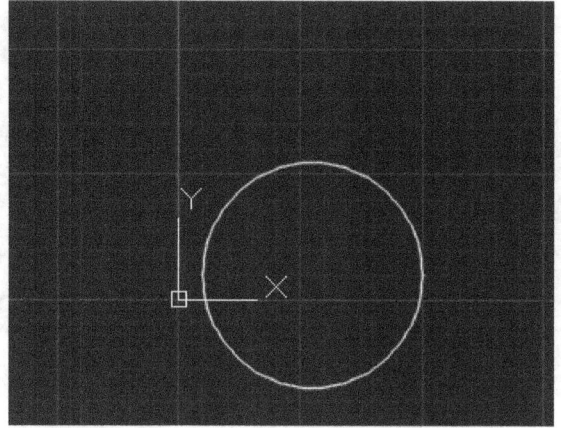

Pic 2.34 Circle created by specifying two points

You can also draw a circle by using 2 tangent and radius. See example below:

1. For example, there are 2 the arcs and I want to draw a circle that tangents to those the arcs and certain radius.

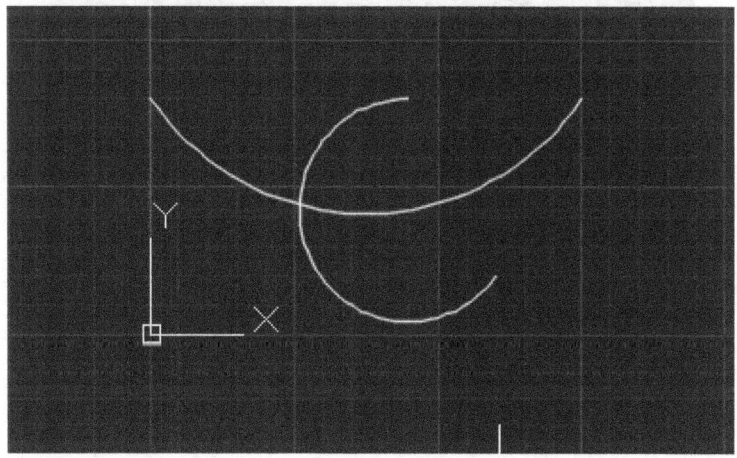

Pic 2.35 Two the arcs

2. Insert circle command and type T in circle parameter.

```
Command: CIRCLE
```

```
Specify center point for circle or [3P/2P/Ttr (tan tan radius)]: t
```

3. Click first the arc.

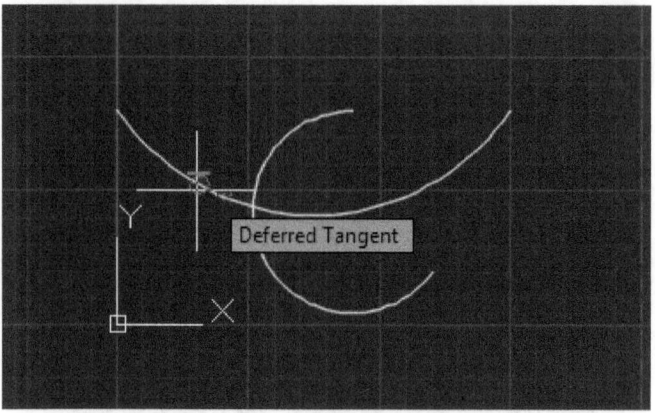

Pic 2.36 Click first the arc

4. Click on the second arc.

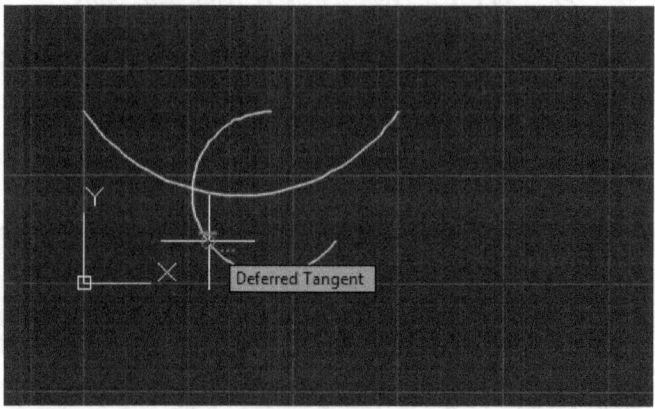

Pic 2.37 Click on second the arc

5. Specify the radius, for this example, I use 50.

```
Command: CIRCLE
Specify center point for circle or [3P/2P/Ttr (tan tan radius)]: t
Specify point on object for first tangent of circle:
Specify point on object for second tangent of circle:
Specify radius of circle <50.0000>: 50
```

6. Click Enter, the circle will be created that tangent to those two the arcs.

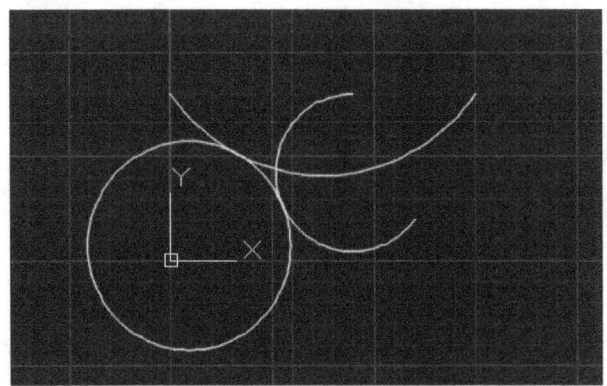

Pic 2.38 Circle with Tan-tan radius

Forth method to draw a circle is by specifying three tangent. See the example below:

1. For example, there are three lines like picture below:

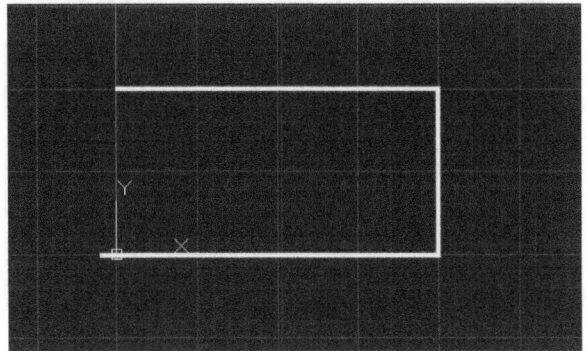

Pic 2.39 Three lines as tangents

2. Click on **Circle > Tan, Tan, Tan** menu.

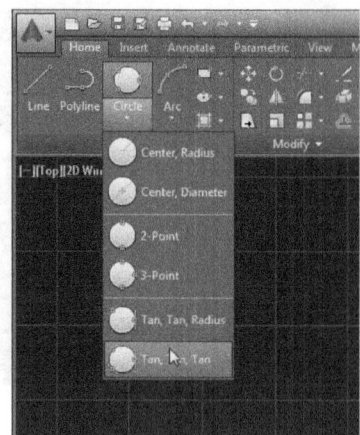

Pic 2.40 Menu Circle > Tan, Tan, Tan

3. Click on the first line.

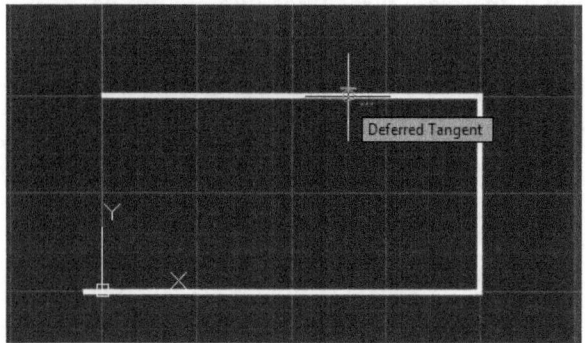

Pic 2.41 Click on the first line

4. Click on the second line.

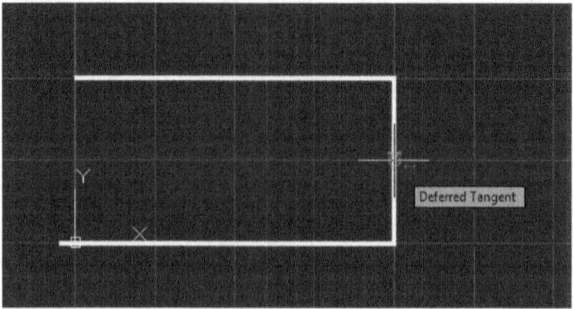

Pic 2.42 Click on second line

5. Click on the third line.

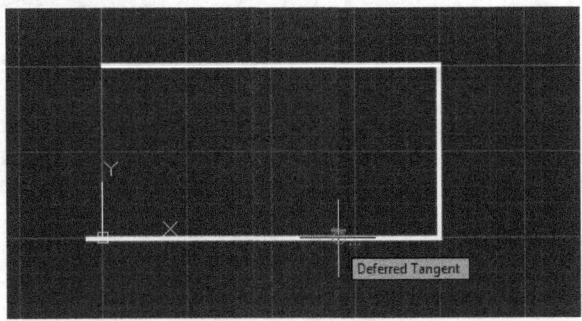

Pic 2.43 Click on the third line

6. The circle will be created.

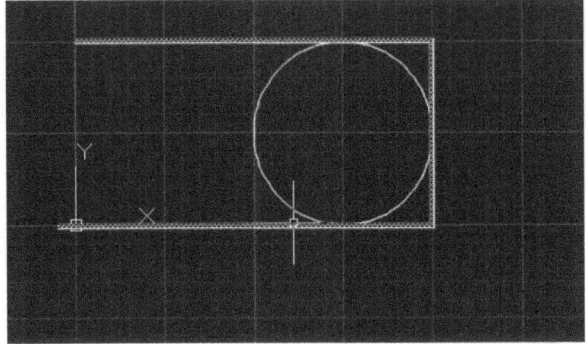

Pic 2.44 The circle created

2.1.4 Draw The arc

The arc can be created using some methods. First by specifying three points. See tutorial below:

1. Type "the arc" in command line.

2. Specify start point to 0,0.

3. Specify second point to 100,50.

4. Specify the third point to 150,0.

```
Command: THE ARC
Specify start point of the arc or [Center]: 0,0
Specify second point of the arc or [Center/End]: 100,50
Specify end point of the arc: 150,0
```

5. The arc will be created:

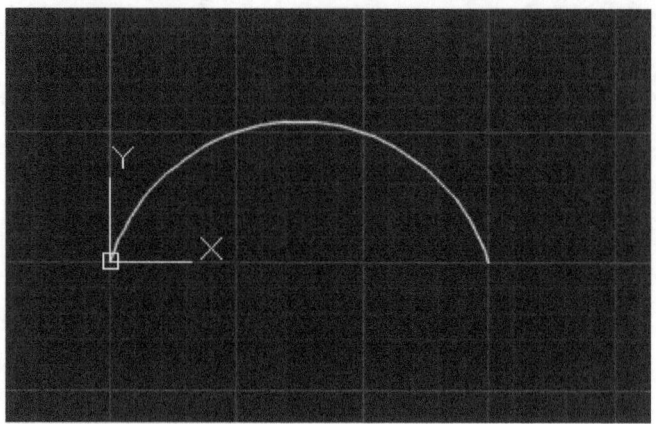

Pic 2.45 The arc created

The second method of creating the arc is by specifying start, center and angle. See steps below:

1. Execute "the arc" command.

2. Specify start point to 0,0.

3. Specify center point of the arc to 50,0.

4. Choose angle and set to -45 degrees.

```
Command: THE ARC
Specify start point of the arc or [Center]: 0,0
Specify second point of the arc or [Center/End]: C
Specify center point of the arc: 50,0
Specify end point of the arc (hold Ctrl to switch direction) or
[Angle/chord Length]: -45
```

5. See picture below for the arc created:

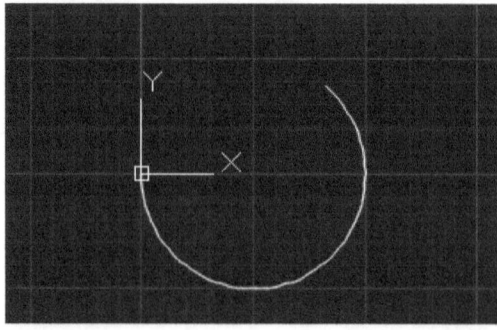

Pic 2.46 The arc created by start point, center, and angle

The second method to create the arc is by specifying Start, Center, Length.

1. Execute the arc command, specify start point of the arc to 0,0.

```
Command: THE ARC
Specify start point of the arc or [Center]: 0,0
Specify second point of the arc or [Center/End]: C
```

2. In Specify second point of the arc, click C to specify Center.

3. Specify center to 50,-40.

4. Insert L to specify length.

5. Specify length = 40.

```
Specify center point of the arc: 50,-40
Specify end point of the arc (hold Ctrl to switch direction) or
[Angle/chord Length]: L
Specify length of chord (hold Ctrl to switch direction): 40
```

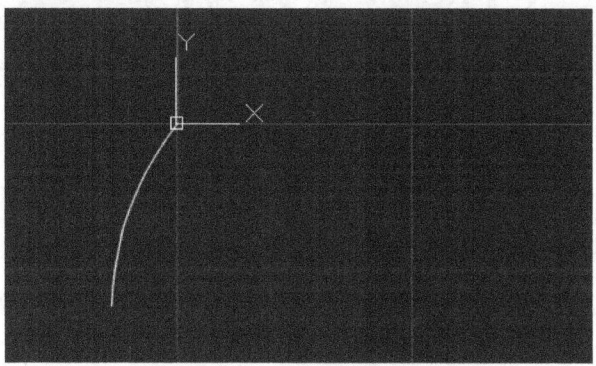

Pic 2.47 The arc created using the arc, center, and length

Next the arc type creation method is by specifying Start End angle. Just enter the start point, end point, and angle. See steps below:

1. Run The arc command, and specify start point to 0,0.

2. Choose E to specify "End point" method.

3. Specify point 100,100 for end point.

```
Command: THE ARC
Specify start point of the arc or [Center]: 0,0
Specify second point of the arc or [Center/End]: E
Specify end point of the arc: 100,100
```
4. Insert A to specify Angle.

5. Type -30 for the angle.

```
Specify center point of the arc (hold Ctrl to switch direction) or
[Angle/Direction/Radius]: A
Specify included angle (hold Ctrl to switch direction): -30
```

6. The arc will be created.

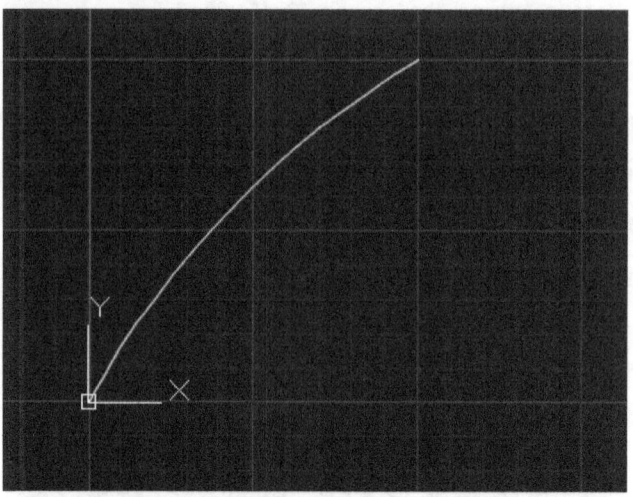

Pic 2.48 The arc created using Start End, and Angle

Next the arc type is Start End, Direction. Here's how to create it:

1. Execute the arc command.

2. Specify start point to 0,0.and type E to select **End** method.

3. Specify End point to 100,0.

```
Command: THE ARC
Specify start point of the arc or [Center]: 0,0
Specify second point of the arc or [Center/End]: E
Specify end point of the arc: 100,0
```

4. Choose D for direction.

```
Specify center point of the arc (hold Ctrl to switch direction) or
[Angle/Direction/Radius]: D
```

5. Specify tangent direction to -45.

```
Specify tangent direction for the start point of the arc (hold Ctrl
to switch direction): -45
```

6. The arc will be created:

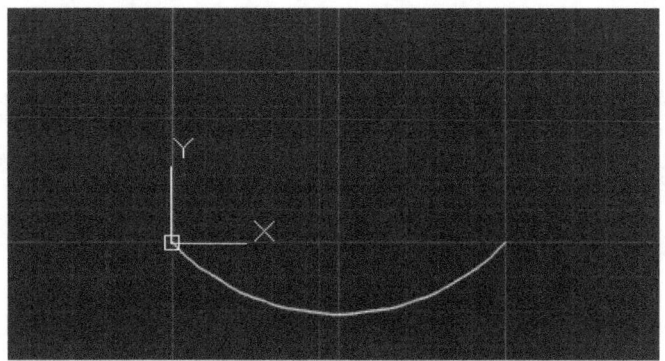

Pic 2.49 The arc created using Start, End, Direction method

Another method to create the arc is by using Start End Radius method. See steps below:

1. Click on The arc > Start, End, Radius.

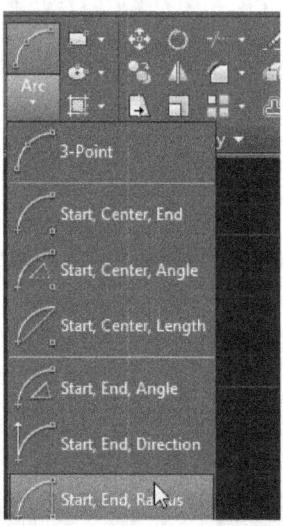

Pic 2.50 Start, End, Radius menu

2. This will set the Start, End, Radius method to create the arc.

3. Specify start point of the arc to 0,0.

```
Command: _the arc
Specify start point of the arc or [Center]: 0,0
Specify second point of the arc or [Center/End]: _e
```

4. Specify end point of the arc to 100,100.

```
Specify end point of the arc: 100,100
```

5. Specify radius to 90.

```
Specify center point of the arc (hold Ctrl to switch direction) or
[Angle/Direction/Radius]: _r
Specify radius of the arc (hold Ctrl to switch direction): 90
```

6. You can see the result below:

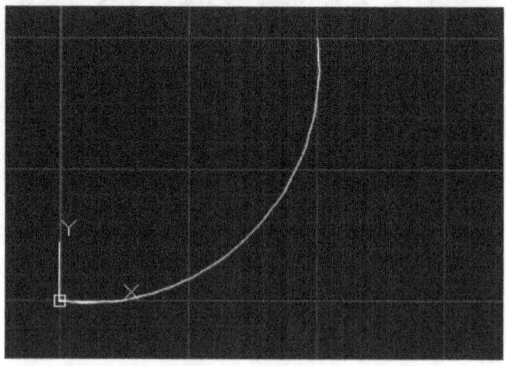

Pic 2.51 The arc created

You can also use Center, Start, End method to create the arc. See steps below:

1. Insert the arc command, then enter C argument for Center.

```
Command: THE ARC
Specify start point of the arc or [Center]: C
```

2. Specify center of the arc = 50,0, and set start point to 0,0.

```
Specify center point of the arc: 50,0
Specify start point of the arc: 0,0
```

3. Specify end point to 100,100.

```
Specify end point of the arc (hold Ctrl to switch direction) or
[Angle/chord Length]: 100,0
```

4. You can see the result as below:

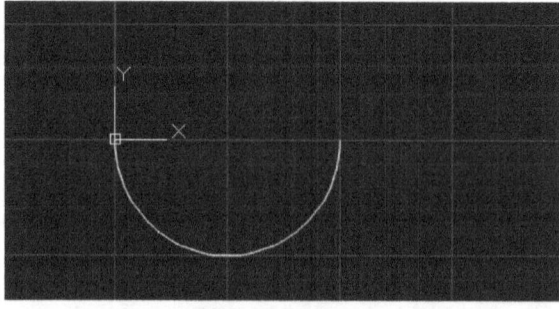

Pic 2.52 Specify center, start, end

Next method is Center Start Angle to create the arc. See steps below:

1. Execute "the arc" command.

2. Choose C for Center.

3. Specify the center point to 50,0.

```
Command: THE ARC
Specify start point of the arc or [Center]: C
Specify center point of the arc: 50,0
```

4. Specify the start point of the arc to 0,0, then choose A for Angle.

```
Specify start point of the arc: 0,0
Specify end point of the arc (hold Ctrl to switch direction) or
[Angle/chord Length]: A
```

5. Specify the angle = 45 degrees.

```
Specify included angle (hold Ctrl to switch direction): 45
```

6. The result will be as below:

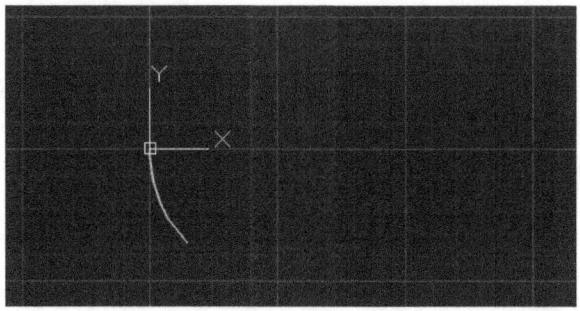

Pic 2.53 The arc creating using "center, start, angle"

Next method is by using Center, Start, and Length. See the steps below:

1. Run "the arc" command.

2. Choose C for Center.

```
Command: THE ARC
Specify start point of the arc or [Center]: C
```

3. Specify the center point of the arc to 50,0.

4. Specify the start point to 0,0.

5. Choose L in Angle/Chord Length.

```
Specify center point of the arc: 50,0
Specify start point of the arc: 0,0
Specify end point of the arc (hold Ctrl to switch direction) or
[Angle/chord Length]: L
```

6. Specify the length of the arc to 100.

```
Specify length of chord (hold Ctrl to switch direction): 100
```

7. The arc will be as below:

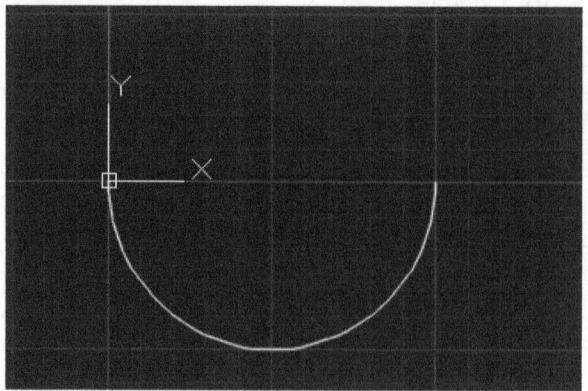

Pic 2.54 The arc already created

To draw a new the arc connected from existing the arc, you can use **Continue**. Here are the steps:

1. After creating the arc.

2. Click on **arc > Continue** in the **Home > Draw** ribbon.

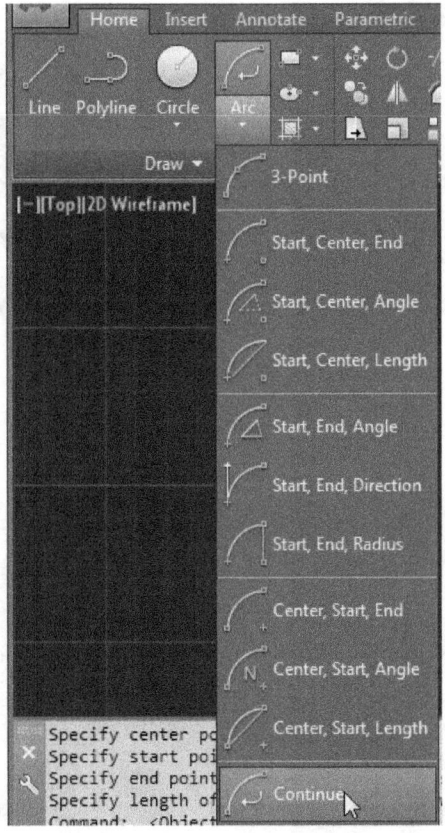

Pic 2.55 Arc > Continue menu

3. You can continue creating the arc from the existing arc.

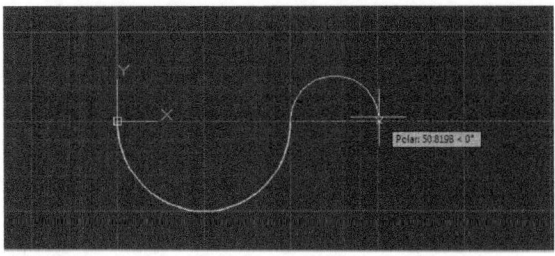

Pic 2.56 Continue creating the arc from the existing arc

4. If you click Enter, another segment of the arc will created.

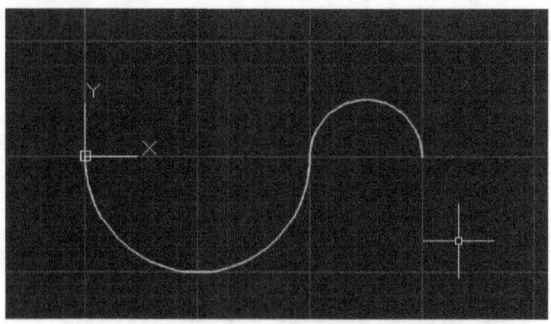

Pic 2.57 The arc segment created

5. By iterating steps above, you can create the arc as much as you want.

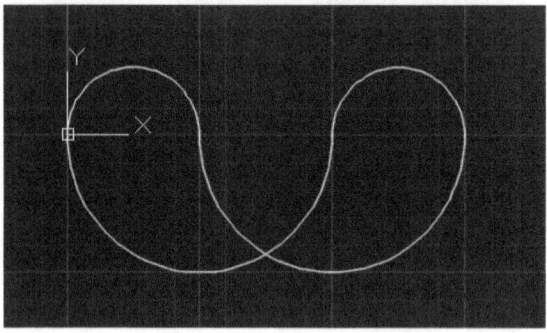

Pic 2.58 The arcs created

2.1.5 Draw Rectangle

To draw rectangle, AutoCAD, use "rectang" command. This will automatically create polyline in rectangle. Just like the arc, there are more than one methods to create rectangle.

Rectang function has many arguments. You can see the example on commands below:

```
Current settings: Rotation = 0
Specify first corner point or [Cham-
fer/Elevation/Fillet/Thickness/Width]:
```

Notes on the arguments:

- First Corner Point: Specifying the first point of rectangle.

- Other Corner Point: Specifying the other point of rectangle.

- Area: Creating rectangle using area, length and width. If Chamfer or Fillet option active, the effect will be appeared on the corner of rectangle.

- Dimensions: Creating rectangle by entering the length and width.

- Rotation: Creating rectangle using certain angle rotation.

- Chamfer: Setting the chamfer of rectangle.

- Elevation: Setting the elevation of rectangle.

- Fillet: Setting fillet radius of rectangle.

- Thickness: Setting the thickness of rectangle.

- Width: Setting the line's width of rectangle.

First tutorial describes how to create simple rectangle, without fillet and width. See steps below:

1. Run "rectang" command.

2. Specify the first corner to 0,0.

```
Command: RECTANG
Specify first corner point or [Cham-
fer/Elevation/Fillet/Thickness/Width]: 0,0
```

3. Specify other corner to 75,0.

```
Specify other corner point or [Area/Dimensions/Rotation]: 75,50
```

4. A simple rectangle will be created on the drawing area.

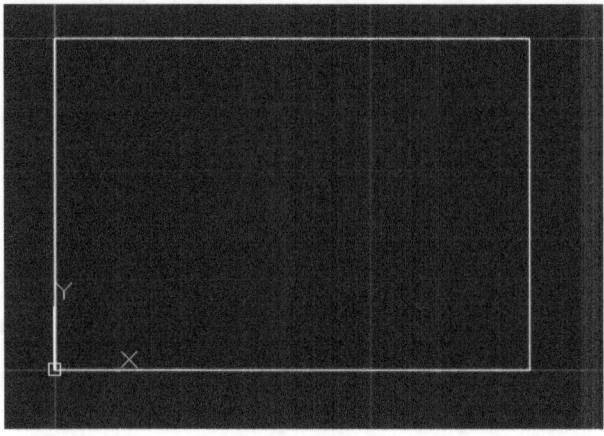

Pic 2.59 Rectangle created using RECTANG command

The rectangle may has a chamfer. See steps below to create the rectangle with a chamfer.

1. Run "RECTANG" command.

2. Choose C on "Specify first corner" to activate the chamfer.

```
Command: RECTANG
Specify first corner point or [Cham-
fer/Elevation/Fillet/Thickness/Width]: C
```

3. On **Specify first chamfer distance**, set to 3, on **specify second chamfer distance**, set to 3.

```
Specify first chamfer distance for rectangles <0.0000>: 3
Specify second chamfer distance for rectangles <3.0000>: 3
```

4. Specify first corner point = 0,0. Then Specify second corner point to 50,50.

```
Specify first corner point or [Cham-
fer/Elevation/Fillet/Thickness/Width]: 0,0
Specify other corner point or [Area/Dimensions/Rotation]: 50,50
```

5. The result is a rectangle with a chamfer on the corner. See picture below:

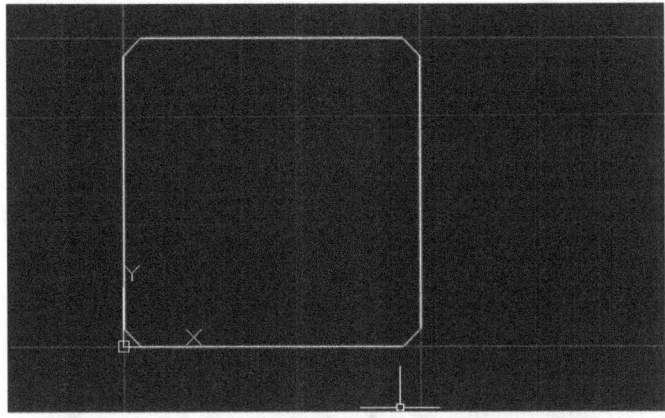

Pic 2.60 a rectangle with chamfer

You can also create fillets on the corner of rectangle. See steps below:

1. Run "rectang" command.

```
Command: RECTANG
```

2. On "Specify first corner", click F.

3. Specify the fillet radius to 3.

4. Then specify first corner point = 0,0.

5. Then Specify another corner point = 50,50.

```
Specify first corner point or [Cham-
fer/Elevation/Fillet/Thickness/Width]: F
Specify fillet radius for rectangles <3.0000>: 3
Specify first corner point or [Cham-
fer/Elevation/Fillet/Thickness/Width]: 0,0
Specify other corner point or [Area/Dimensions/Rotation]: 50,50
```

6. The result is a rectangle with fillet:

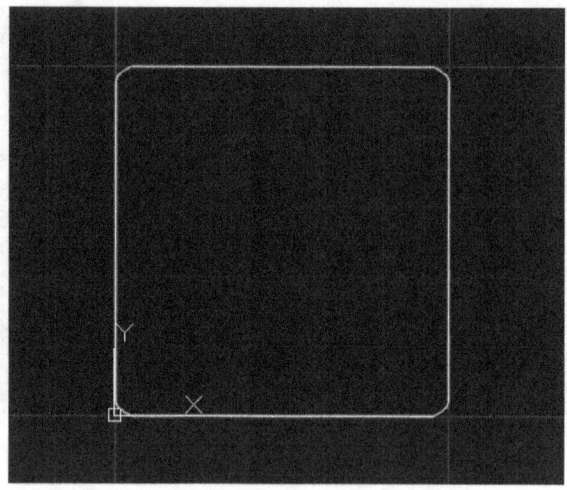

Pic 2.61 A rectangle with fillet

You can also change the width argument of a rectangle to draw a rectangle with custom width. The steps are:

1. Run "RECTANG" command.

```
Command: RECTANG
Current rectangle modes:  Fillet=3.0000
```

2. On "Chamfer/Elevation/Fillet", click w.

3. Set the width to 1.

```
Specify first corner point or [Cham-
fer/Elevation/Fillet/Thickness/Width]: w
Specify line width for rectangles <0.0000>: 1
```

4. Specify the first corner point = 0,0. And then the second corner point =100,50.

```
Specify first corner point or [Cham-
fer/Elevation/Fillet/Thickness/Width]: 0,0
Specify other corner point or [Area/Dimensions/Rotation]: 100,50
```

5. Rectangle created will have custom width.

Pic 2.62 Create Rectangle

2.1.6 Draw Polygon

Polygon will draw a polygon with a custom number of sides. The default number of sides is 4, but you can customize. A polygon can be inscribed or circumscribed. See tutorial below to draw polygon:

1. First, draw a circle

2. Specify the center of the circle to 25.25.

3. Specify the circle radius to 25.

```
Command: CIRCLE
Specify center point for circle or [3P/2P/Ttr (tan tan radius)]:
25,25
Specify radius of circle or [Diameter]: 25
```

4. A circle will be created with center = 25.25 and radius = 25. Create the circle to help you to distinguish between inner or outer polygon.

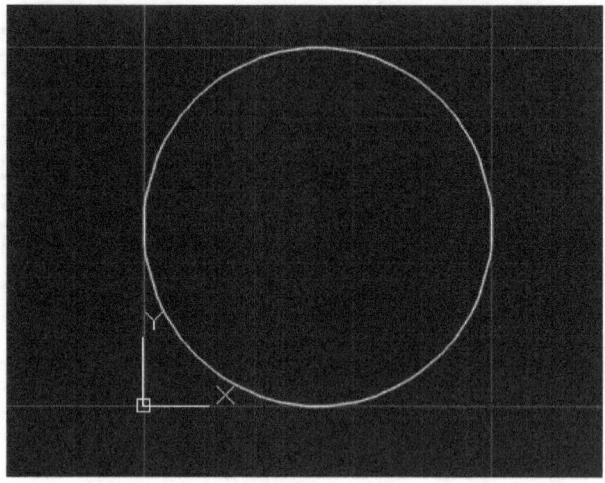

Pic 2.63 Circle created

5. Execute "polygon" command.

6. Enter the number of sides to 5.

```
Command: POLYGON
Enter number of sides <4>: 5
```

7. Specify the center of polygon to 25,25.

8. For first polygon, I choose the inscribed by typing I

```
Specify center of polygon or [Edge]: 25,25
Enter an option [Inscribed in circle/Circumscribed about circle]
<I>: i
```

9. Specify the radius to 25.

```
Specify radius of circle: 25
```

10.   You can see the polygon created, but inscribed in the circle.

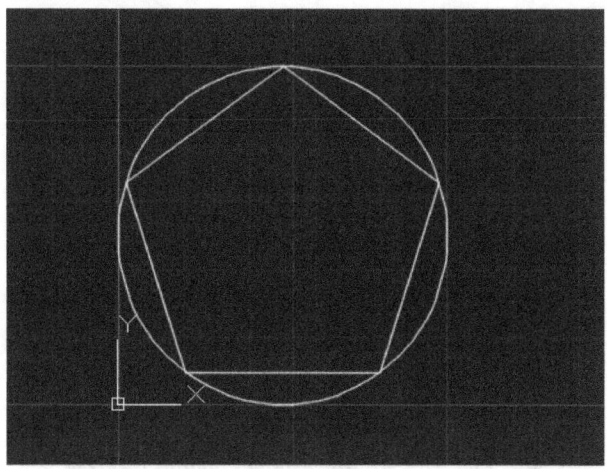

Pic 2.64 Polygon inserted inside the circle

If you want to create circumscribed polygon. Use steps below:

1. Type "polygon" in the command prompt

2. Specify the center of to 25.25.

```
Command: POLYGON
Enter number of sides <5>:
Specify center of polygon or [Edge]: 25,25
```

3. Type C to specify circumscribed.

```
Enter an option [Inscribed in circle/Circumscribed about circle]
<I>: C
```

4. Define radius of circle = 25.

```
Specify radius of circle: 25
```

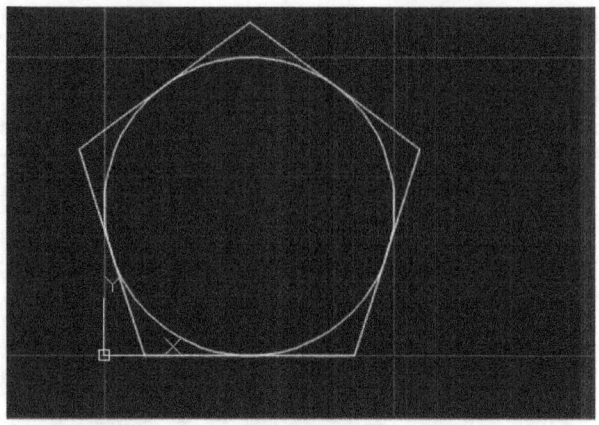

Pic 2.65 The circumscribed polygon inserted

5. You can compare the inscribed polygon yang di dalam atau di luar lingkaran seperti berikut ini:

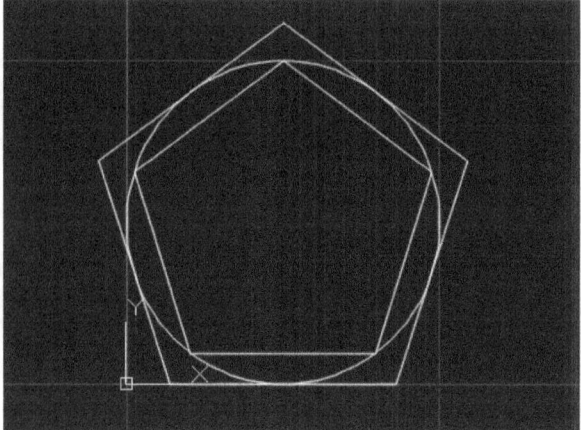

Pic 2.66 Polygons created

2.1.7 Draw Ellipse

To draw ellipse, you have to define the long axis and short axis, see picture below for the details

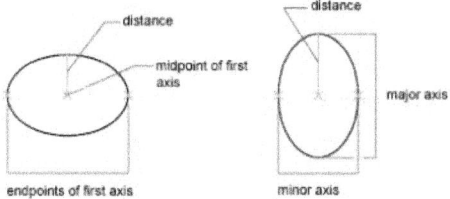

Pic 2.67 Ellipse drawing

Below are steps to create ellipse:

1. Type "ellipse" to create ellipse.

2. Specify axis' end point to 100,50.

3. Specify other endpoint of axis to 0,50.

4. Specify the distance to 20.

```
Command: ELLIPSE
Specify axis endpoint of ellipse or [The arc/Center]: 100,50
Specify other endpoint of axis: 0,50
Specify distance to other axis or [Rotation]: 20
```

5. The ellipse will be created.

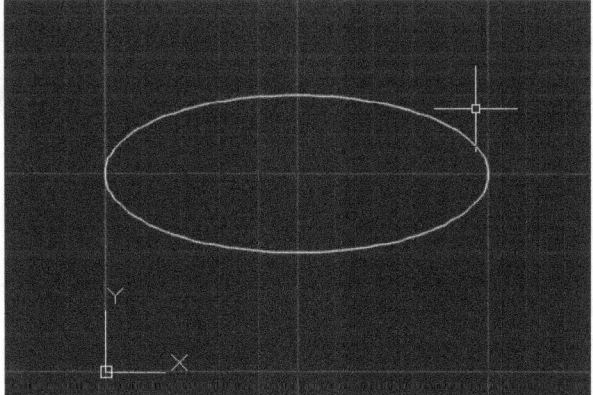

Pic 2.68 Ellipse created

You can create the arc from ellipse. See steps below:

1. Execute "ellipse" command.

2. Choose A to specify the arc.

```
Command: ELLIPSE
Specify axis endpoint of ellipse or [The arc/Center]: A
```

3. Specify axis' end point to 100,0.

4. Specify axis' other end point to 0,0.

```
Specify axis endpoint of elliptical the arc or [Center]: 100,0
Specify other endpoint of axis: 0,0
```

5. Specify distance to other axis = 30 to create the ellipse.

```
Specify distance to other axis or [Rotation]: 30
```

6. Specify start angle to 50 and end angle to 10.

```
Specify start angle or [Parameter]: 50
Specify end angle or [Parameter/Included angle]: 10
```

7. Ellipse the arc will be created.

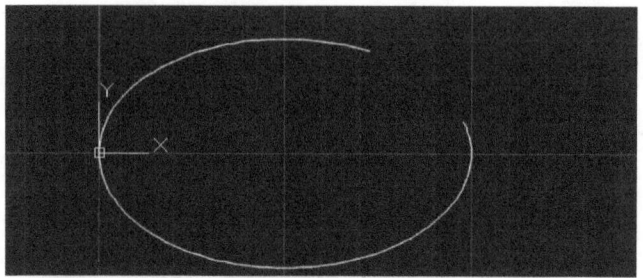

Pic 2.69 Ellipse the arc created

You can also create rotated ellipse, see steps below:

1. Type "ellipse" in command line.

2. Specify ellipse's end point to 100,0.

3. Specify ellipse's other end point to 0,50.

```
Command: ELLIPSE
Specify axis endpoint of ellipse or [The arc/Center]: 100,0
Specify other endpoint of axis: 0,50
```

4. Specify distance to 45.

```
Specify distance to other axis or [Rotation]: 45
```

5. Specify rotation to 45.

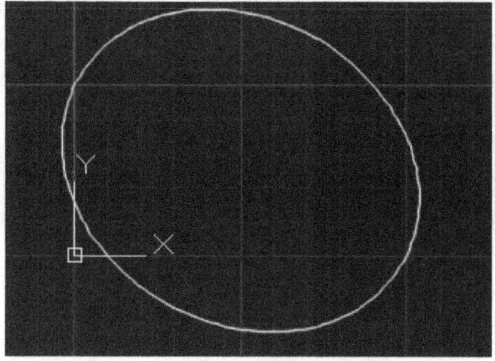

Pic 2.70 Rotated the arc created

2.1.8 Draw Hatch

Certain area can be hatched, you can also define the hatch type, see example below for drawing hatch:

1. Create two objects. One circle and one polygon with number of sides = 5.

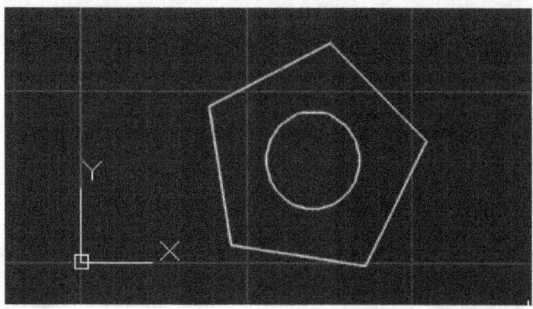

Pic 2.71 Creating two objects, circle and polygon

2. Select both objects by your mouse.

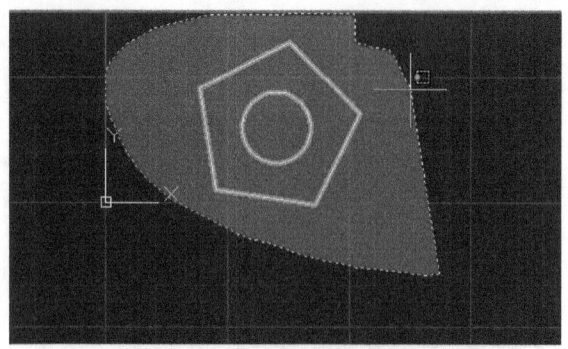

Pic 2.72 Selecting objects

3. Both objects selected, see following pic:

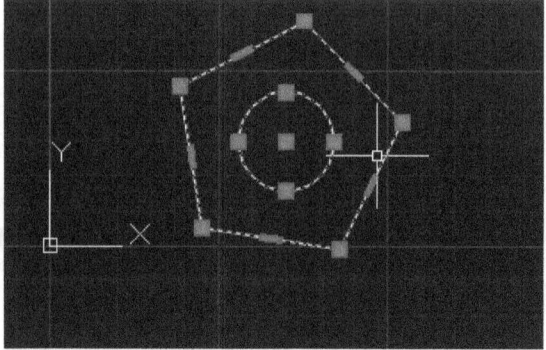

Pic 2.73 Both objects selected

4. Right click until the context menu appears, and select **Group > Group**.

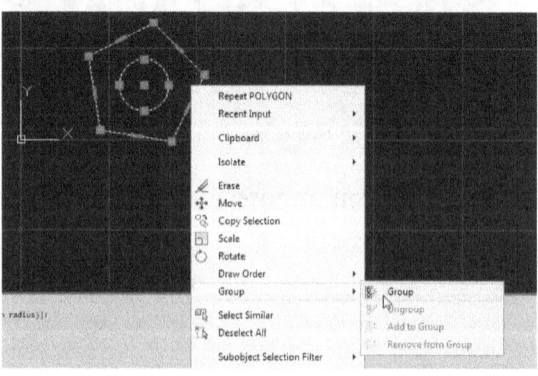

5. Objects will be grouped, then choose the object.

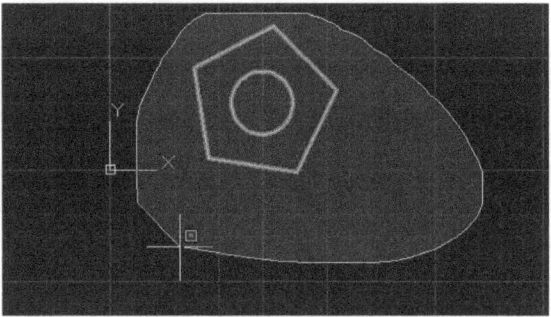

Pic 2.75 Select the grouped object

6. After selected, you can see the objects become one entity (because it's already grouped using Group > Group menu).

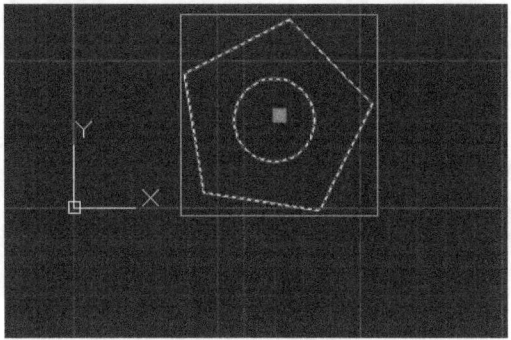

Pic 2.76 Grouped object

7. To hatch area between circle and polygon, click **Hatch** in **Home > Draw** box.

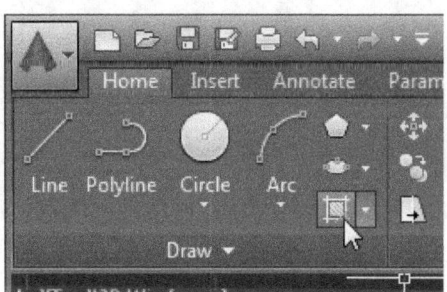

Pic 2.77 Click on Hatch button

8. In Pattern box, click on the arrow button to display more patterns.

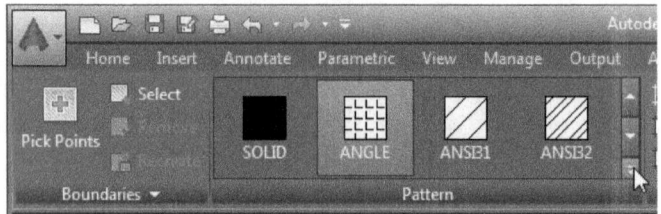

Pic 2.78 Click on arrow to display more patterns

9. You can see list of hatch's pattern.

Pic 2.79 Hatch patterns

10. After selecting the pattern, click on the area.

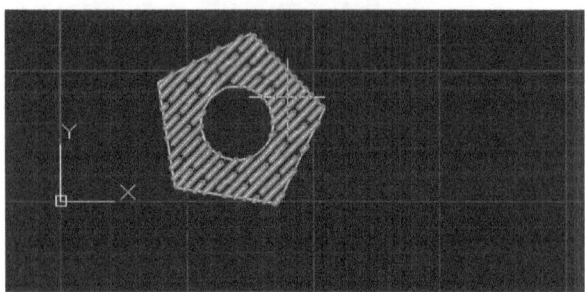

Pic 2.80 Click on the area to be hatched

11. The hatch will be created:

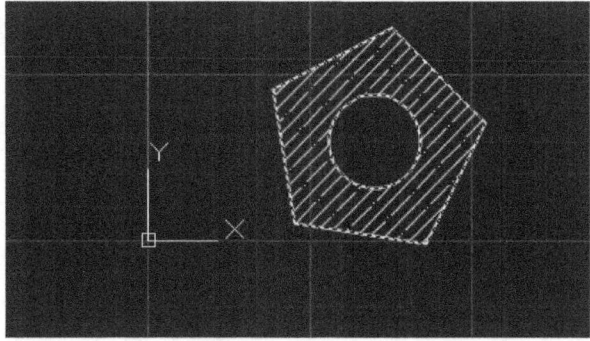

Pic 2.81 Hatched area

2.1.9 Draw Spline

Spline command is used to create curvy line. You can use Fit method or CV method. See tutorial below:

1. Run "spline" command in AutoCAD.

```
Command: SPLINE
Current settings: Method=Fit    Knots=Chord
```

2. Specify the first point to 0,0 and the next point to 25,25.

```
Specify the first point or [Method/Knots/Object]: 0,0
Enter next point or [start Tangency/toLerance]: 25,25
```

3. Specify next point to 50,0 and 75,25

```
Enter next point or [end Tangency/toLerance/Undo]: 50,0
Enter next point or [end Tangency/toLerance/Undo/Close]: 75,25
```

4. Specify next point to 100,0 and 50,-50. Then click C on your keyboard to close spline.

```
Enter next point or [end Tangency/toLerance/Undo/Close]: 100,0
Enter next point or [end Tangency/toLerance/Undo/Close]: 50,-50
Enter next point or [end Tangency/toLerance/Undo/Close]: C
```

5. See following picture to see the spline result:

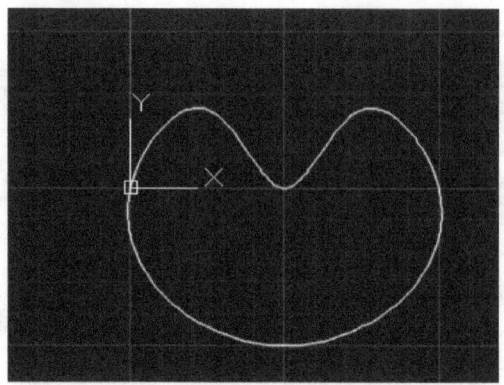

Pic 2.82 Spline result

Spline can also use CV method. See steps below:

1. Enter "spline" command and click "m" on your keyboard to specify Method.

```
Command: SPLINE
Specify the first point or [Method/Degree/Object]: m
```

2. Insert cv to choose cv method for spline creation.

```
Enter spline creation method [Fit/CV] <CV>: cv
Current settings: Method=CV    Degree=3
```

3. Specify the first point to 0,0 and the next point to 25,25.

```
Specify the first point or [Method/Degree/Object]: 0,0
Enter next point: 25,25
```

4. Specify next point to 50,0 and 75,25.

```
Enter next point or [Undo]: 50,0
Enter next point or [Close/Undo]: 75,25
```

5. Insert next point 100,0 and click C butotn on your keyboard to close the spline.

```
Enter next point or [Close/Undo]: 100,0
Enter next point or [Close/Undo]: C
```

6. See following pic for the result.

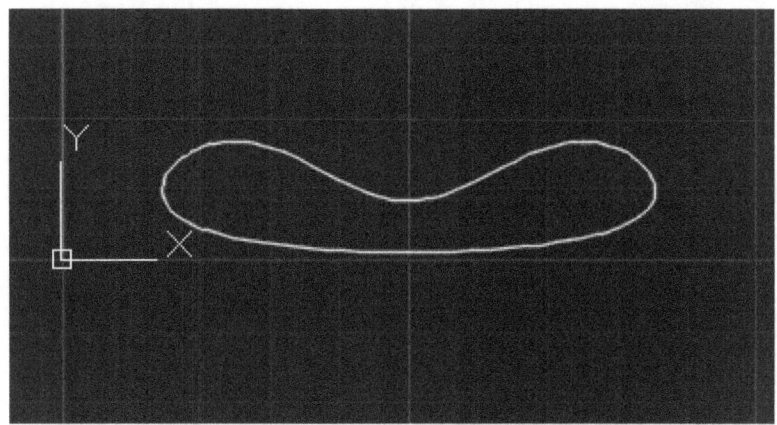

Pic 2.83 Spline result created with cv method

2.1.10 Draw XLINE

Xline is infinite line, commonly used in construction drawing. Xline command enables you to create infinite line just by specifying two points.

See tutorial below to draw XLINE:

1. Execute "xline" command.

2. Specify the first point to 50,0.

3. Specify second point to 50,10.

```
Command: XLINE
Specify a point or [Hor/Ver/Ang/Bisect/Offset]: 50,0
Specify through point: 50,10
```

4. An infinite vertical line that passes two points specified will be created.

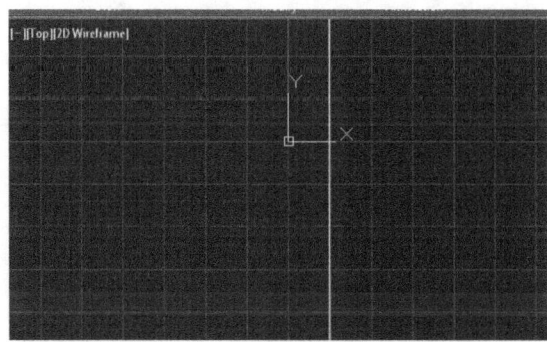

5. To create infinite horizontal line, type "xline".

6. Set the first point to 50,50 and the second point to 100,50.

```
Command: XLINE
Specify a point or [Hor/Ver/Ang/Bisect/Offset]: 50,50
Specify through point: 100,50
```

7. A horizontal xline will be created that passes the two points.

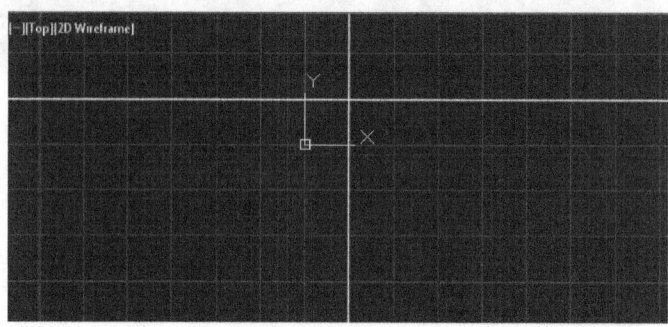

Pic 2.85 Horizontal xline created

8. To create xline with specified angle, first execute "xline".

9. Choose "a" for Angle by clicking "A" button on your keyboard.

10. Set angle to 30 degrees.

11. Specify point to 50,50.

```
Command: XLINE
Specify a point or [Hor/Ver/Ang/Bisect/Offset]: a
Enter angle of xline (0) or [Reference]:  30
Specify through point: 50,50
```

12. See following pic for the result.

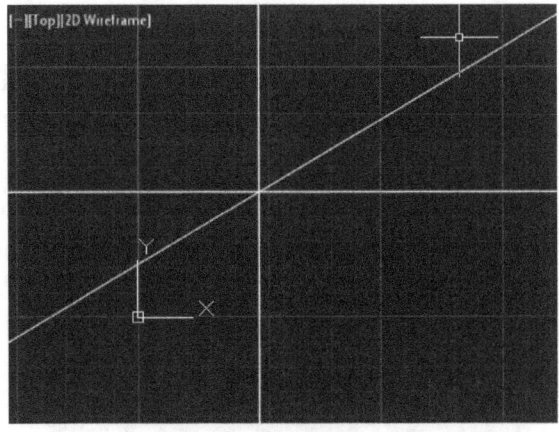

Pic 2.86 Xline with angle

2.1.11 Draw RAY

Ray similar with xline, but ray have start point. See tutorial below for creating ray line:

1. Type "ray" for Ray line.

2. Specify start point to 50,50.

3. Specify the through point to 75,75 and 100,50 and 100,25.

```
Command: _ray Specify start point: 50,50
Specify through point: 75,75
Specify through point: 100,50
Specify through point: 100,25
```

4. The result will be as below:

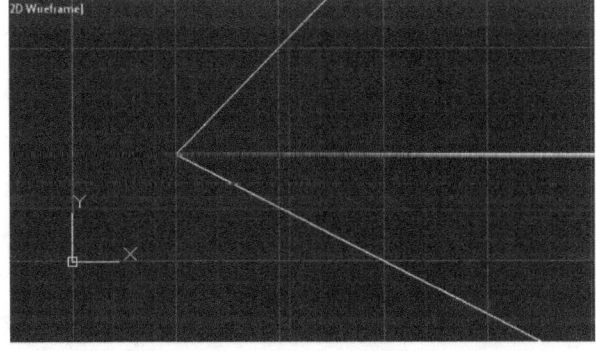

Pic 2.87 Creating line with Ray command

2.1.12 Divide

DIVIDE command divides line or object to some segments. This is suitable for creating dimension's annotation . See tutorial below for the detail:

1. For example there is a line I want to divide into segments.

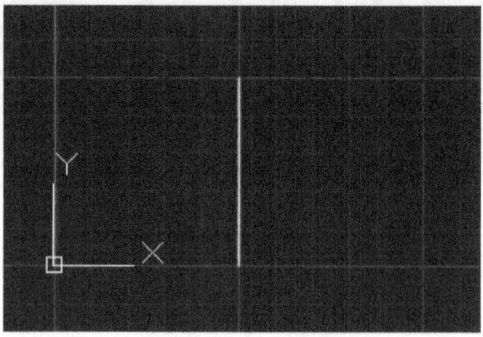

Pic 2.88 A line to divide

2. Type "divide" or click Divide button in **Home > Draw**.

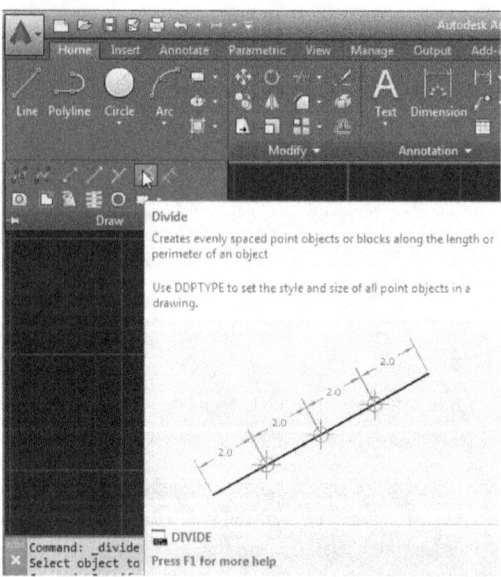

Pic 2.89 Click Divide button

3. Select this line.

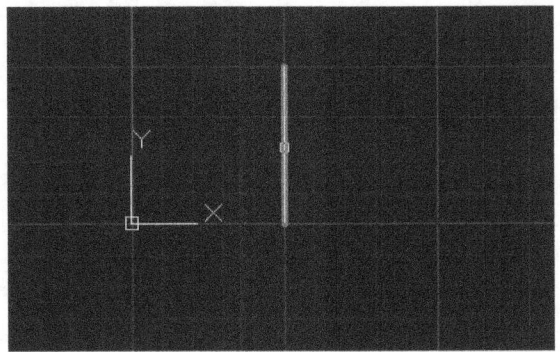

Pic 2.90 Click on the line

4. Select the line you want to divide, selected line will become dotted line.

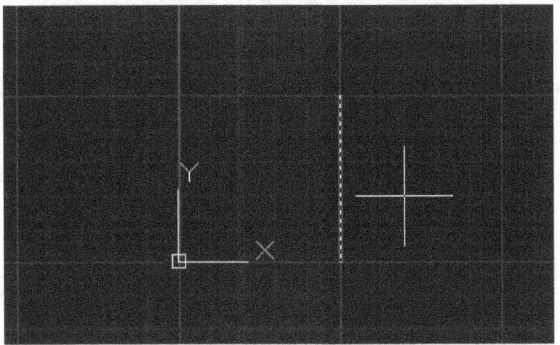

Pic 2.91 Selected Line become dotted line

5. Enter number of segment to 5, this will create 5 segments or 4 points inside the line.

```
Command: _divide
Select object to divide:
Enter the number of segments or [Block]: 5
```

6. If the line moved, you can see 4 points, the points were created by "divide" command.

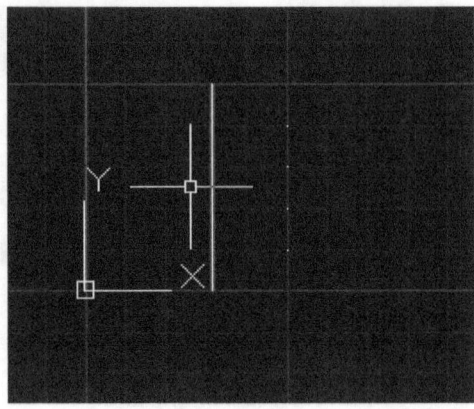

Pic 2.92 Four points that divide line to 5 segments already created

2.1.13 Draw Helix

Helix command creates helix object. You just have to specify the lower diameter, upper diameter and the height. See example below:

1. Type "helix" in the command prompt.

```
Command: HELIX
Number of turns = 3.0000    Twist=CCW
```

2. Specify center point of the base to 50,50. Then specify base's radius to 30 and top's radius to 30. In this example, I use same radius for top and base.

3. Specify the height to 50.

```
Specify center point of base: 50,50
Specify base radius or [Diameter] <22.3607>: 30
Specify top radius or [Diameter] <30.0000>: 30
Specify helix height or [Axis endpoint/Turns/turn Height/tWist]
<50.9902>: 50
```

4. The helix will be created.

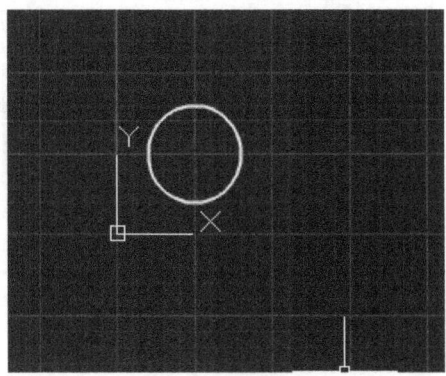

Pic 2.93 Helix created

5. What you see is circle because helix is 3d object, and you only see from the top. To view from side, change the WCS navigator.

Pic 2.94 Changing the WCS navigator

6. Change the view to front.

Pic 2.95 Changing WCS view to front

7. See following pic for the helix seen from side view.

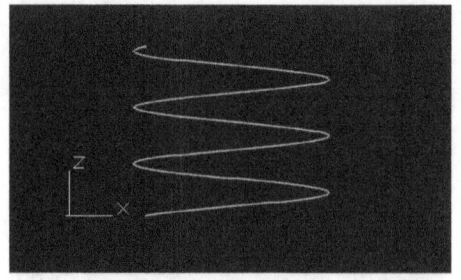

Pic 2.96 Helix seen in side view

8. Back again to TOP view.

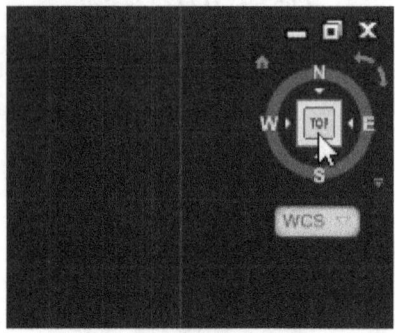

Pic 2.97 Returning to TOP view

9. The helix will be circle again, because the base radius = top radius.

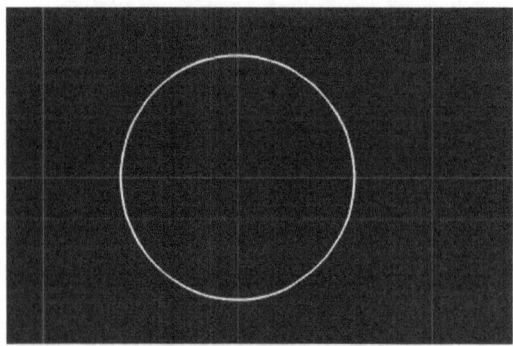

Pic 2.98 Helix seen from above

2.1.14 Draw Donut

Donut command used to create object similar to donut, that is a circle with inside diameter, and outside diameter. See following example:

1. Type "donut" command.

2. Set inside diameter to 50 and outside diameter to 70.

```
Command: DONUT
Specify inside diameter of donut <50.0000>: 50
Specify outside diameter of donut <70.0000>: 70
```

3. Specify center coordinate to 50,50.

```
Specify center of donut or <exit>: 50,50
```

4. See the result in the following pic.

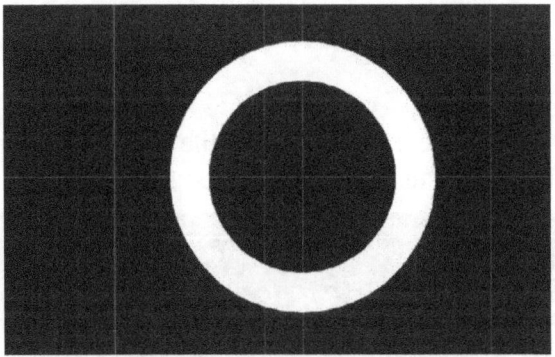

Pic 2.99 Donut created

2.2 Modify 2D Drawing

The 2D Drawing already created, can be modified again. AutoCAD has lots of functions to accommodate modification.

2.2.1 Move

Move command used to move existing object to other place. It's common to use relative coordinate or polar coordinate to move the object. See example below:

1. For example, I have object like this.

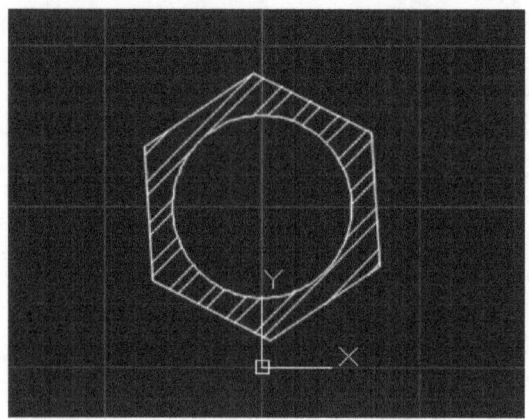

Pic 2.100 Object that will be moved

2. Type "move", then choose object you want to move.

```
Command: MOVE
Select objects: 3 found, 1 group
Select objects: click object
```

3. Click on the object, and select the base point by inserting coordinate or click using your mouse.

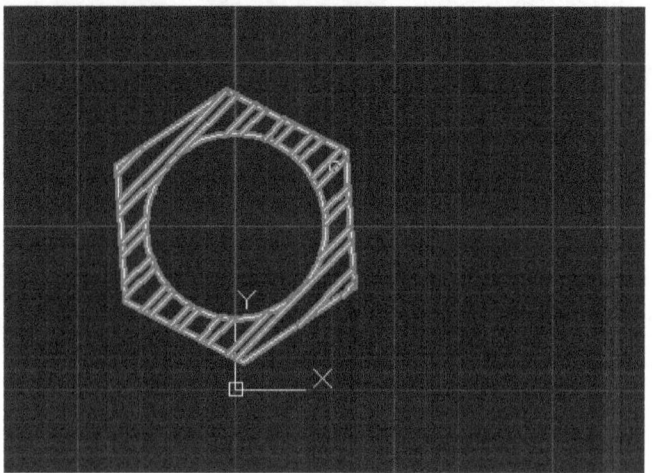

Pic 2.101 Click on object

4. Specify the base point.

```
Specify base point or [Displacement] <Displacement>: click
```

5. For example, I use center of my circle as base point.

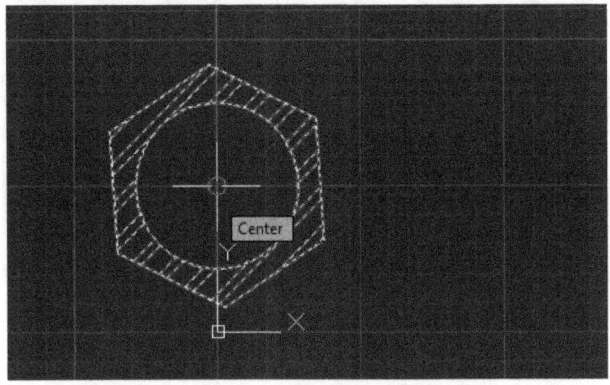

Pic 2.102 Click on object's center

6. Specify second point where you want the the base coordinate to be moved into.

```
Specify second point or <use the first point as displacement>:
```

7. When you want to click, you can see the preview of the object.

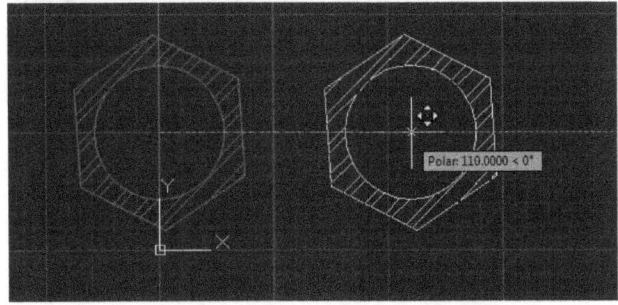

Pic 2.103 Initial and final position

8. Click Enter, the object will have new position.

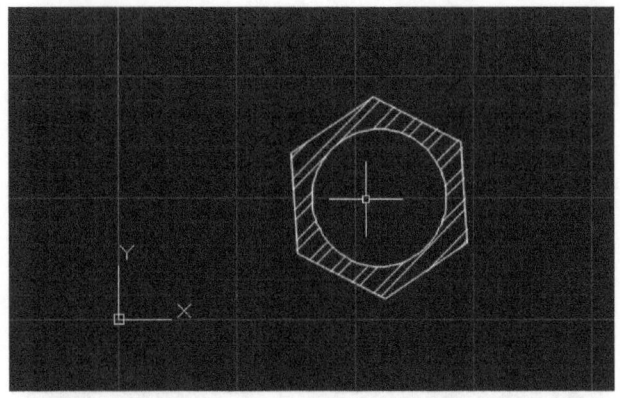

Pic 2.104 New object's position

2.2.2 Rotate

Rotate command rotates object based on base point and degrees of rotation. See example below:

1. For example, I have object on following pic:

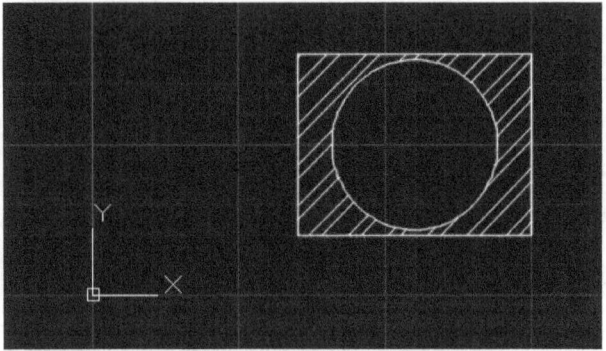

Pic 2.105 Object to rotate

2. Type rotate, then select object you want to rotate.

```
Command: ROTATE
Current positive angle in UCS:  ANGDIR=counterclockwise  ANGBASE=0
Select objects: 3 found, 1 group
Select objects: [click on object]
```

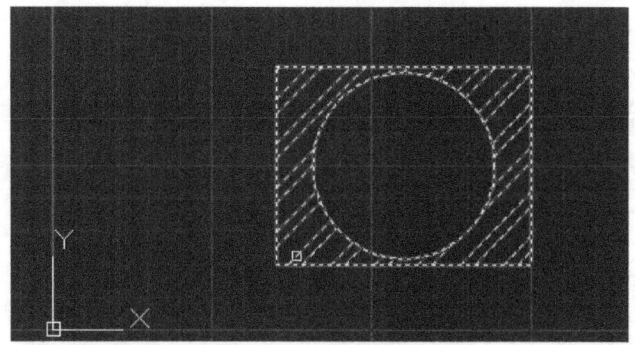

Pic 2.106 Selecting the object

3. Specify the base point for rotation.

```
Specify base point: [click on base point]
```

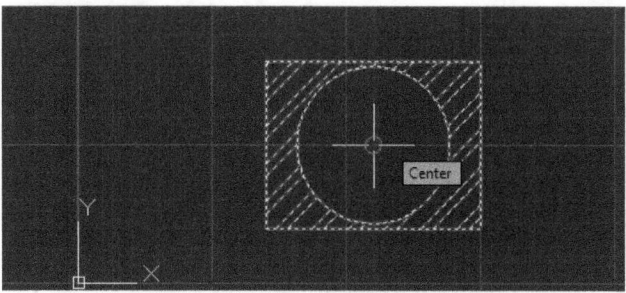

Pic 2.107 Specify the base point

4. If already clicked, rotation icon appears.

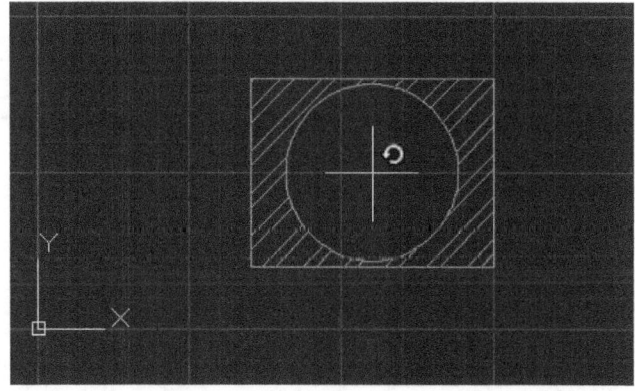

Pic 2.108 Rotation icon appears

5. Set the rotation degrees to -45.

```
Specify rotation angle or [Copy/Reference] <0>: -45
```

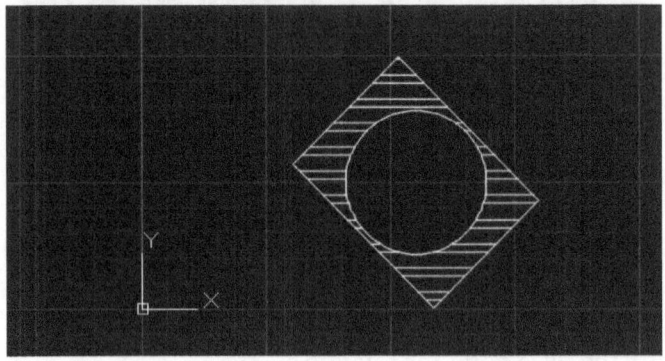

Pic 2.109 Object after rotated to -45 using the center of circle as base point

6. You can also rotate to 90 degrees:

```
Command: ROTATE
Current positive angle in UCS: ANGDIR=counterclockwise ANGBASE=0
Window Lasso  Press Spacebar to cycle options3 found, 1 group
Select objects:
Specify base point:
Specify rotation angle or [Copy/Reference] <45>: 90
```

7. The object will be rotated by 90 degrees.

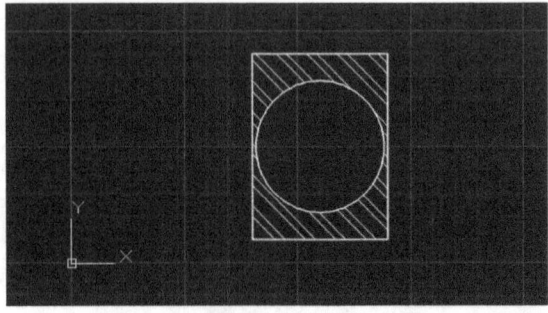

Pic 2.110 Object 90 degrees rotated

2.2.3 Trim

Trim command trims certain part of objects. See steps below to see the example of TRIM function:

1. For example there is three circle objects.

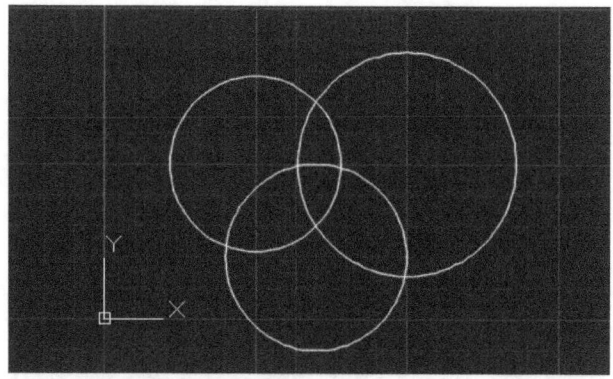

Pic 2.111 Three circle objects

2. You'll trim the inside part of the intersection. Type "trim first".

3. Select all objects.

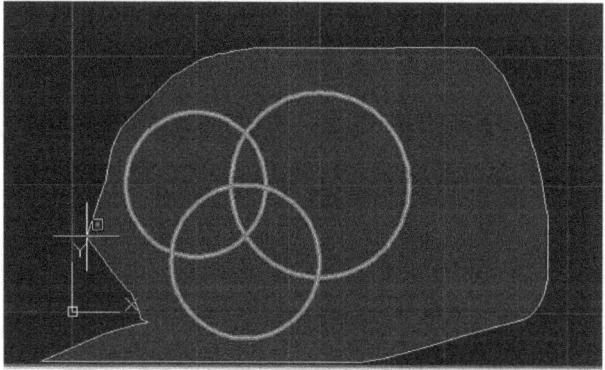

Pic 2.112 Select all objects

4. Selected objects will becoming dotted line.

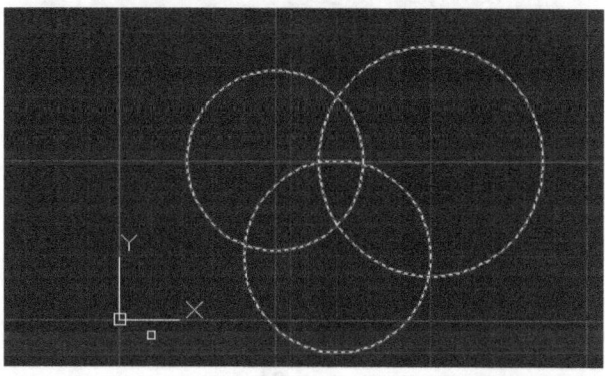

Pic 2.113 Selected objects in dotted line

5. Click on the segments you want to trim.

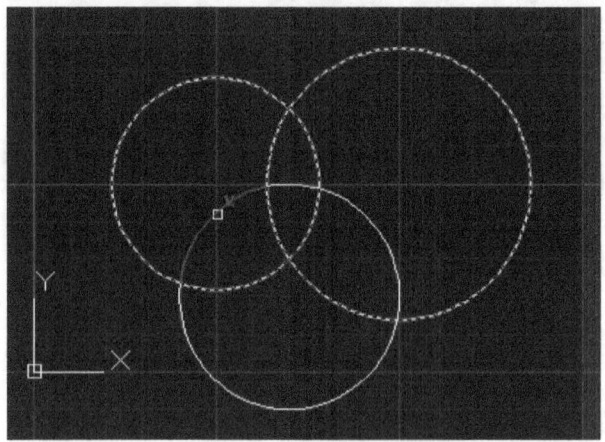

Pic 2.114 Click on segments you want to trim

6. The segment you click will disappear/trimmed. If ERASE erase all of the object, trim will erase selected segment of object.

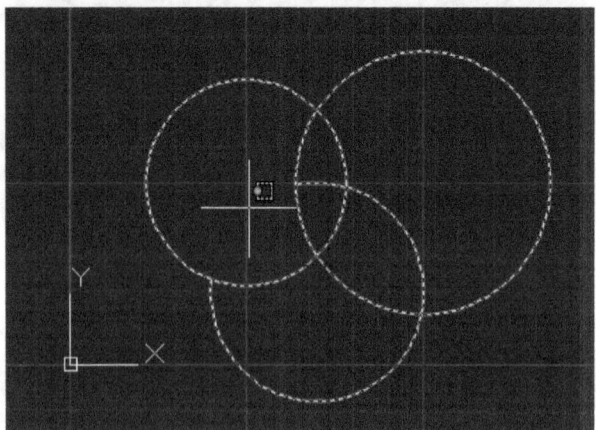

Pic 2.115 Trimmed segment disappears

7. You can click other segment to trim.

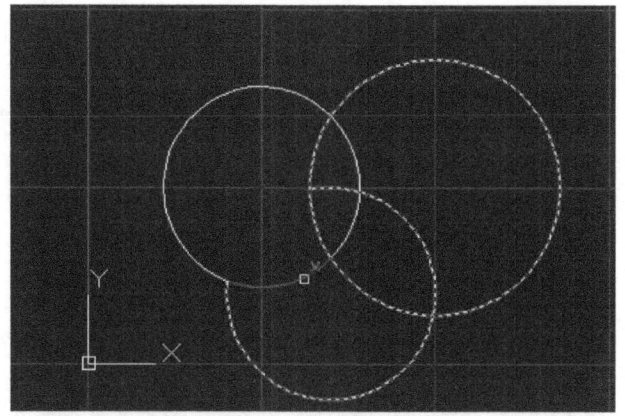

Pic 2.116 Clicking other segment to trim

8. The other segment will diasppear too.

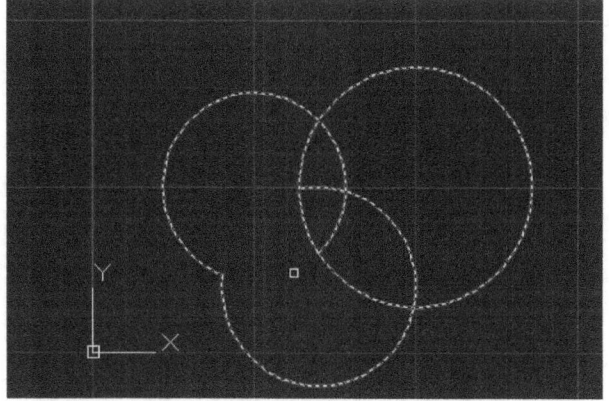

Pic 2.117 Second segment disappears

9. You can click other segment to trim it.

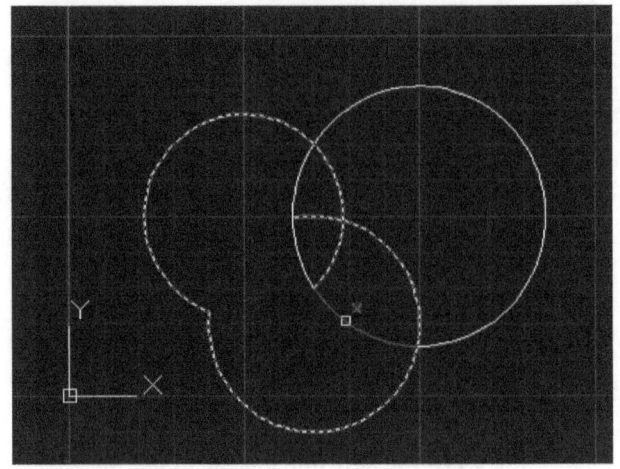

Pic 2.118 Selecting the segment

10.   Final result will be as below:

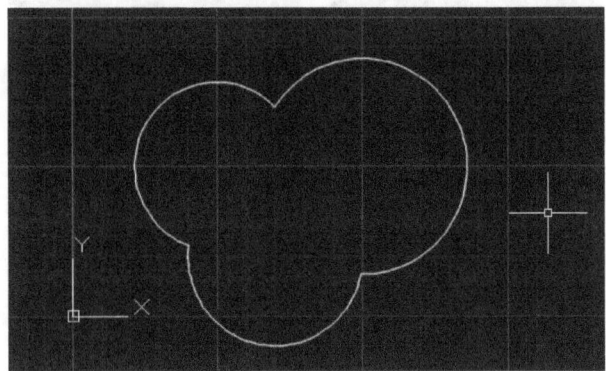

Pic 2.119 Result of trim process

2.2.4 Extend

Extend command extends line or the arc to certain object. See example below for more advanced:

1. For example, there are one the arc and one line. The arc is going to be extended to the line.

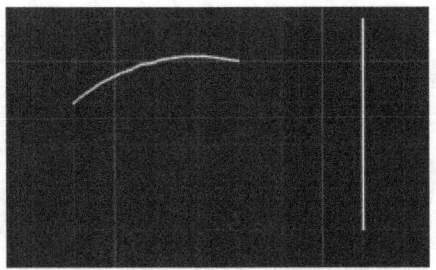

Pic 2.120 An arc and a line

2. Type "extend" command.

3. Select all objects.

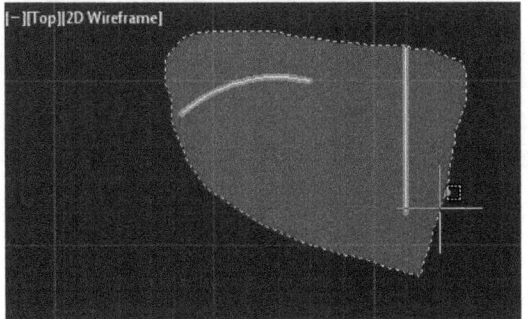

Pic 2.121 Selecting all objects

4. Both objects becoming dotted line.

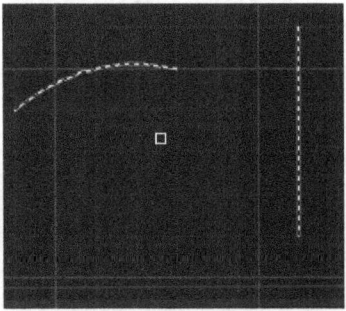

Pic 2.122 Selected objects becoming dotted line

5. Click on object you want to extend, the object become extended.

```
Select object to extend or shift-select to trim or
[Fence/Crossing/Project/Edge/Undo]:
```

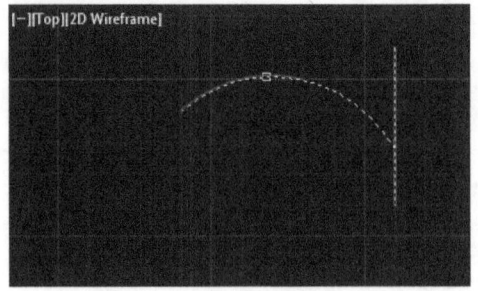

Pic 2.123 Object extended

6. See following pic for the result.

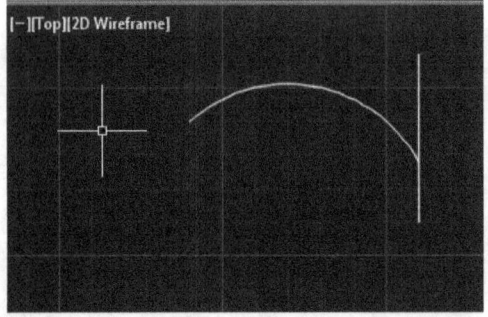

Pic 2.124 The arc has been extended

2.2.5 Erase

Erase command erases selected object. Erase will erase all part of selected object, not only the segments. Here's how to use erase object:

1. From picture below, the hatch will be erased.

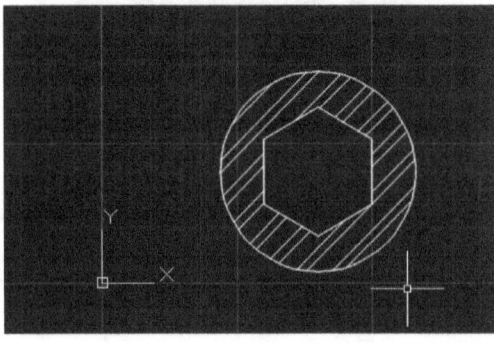

2. Type "erase" in document.

3. Pointer icon will be changed to erase mode.

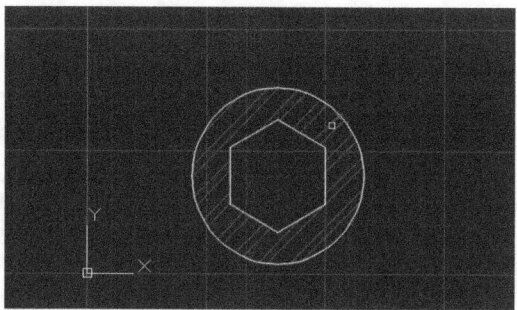

Pic 2.126 Pointer ready to erase

4. Click on the hatch to erase and click **Enter**.

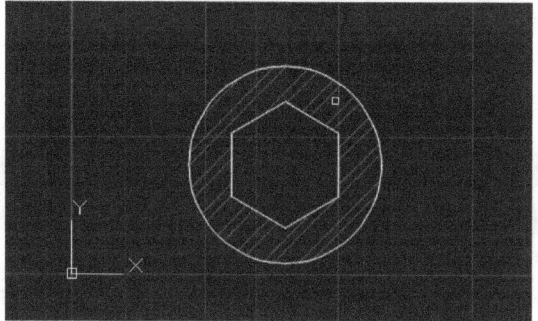

Pic 2.127 Click on the hatch

5. The hatch will be erased.

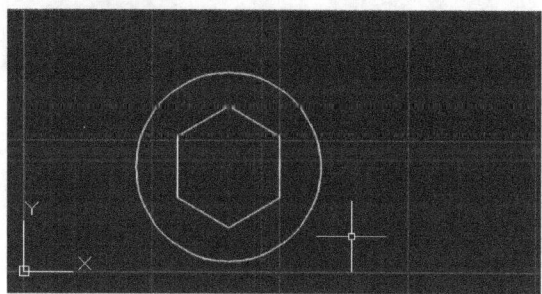

Pic 2.128 Object after the hatch erased

2.2.6 Copy

Copy command copies object, where the copied object still exist. See example below:

1. Type "Copy".

2. Choose the object to copy.

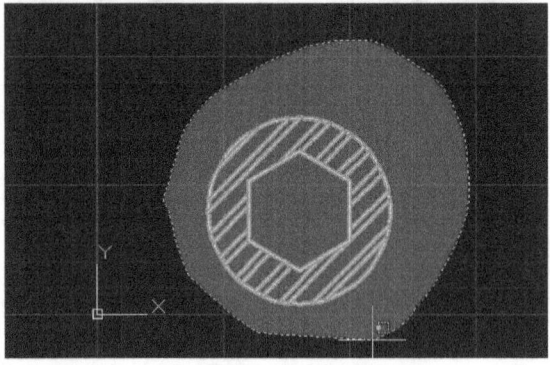

Pic 2.129 Selecting object to copy

3. Select base point.

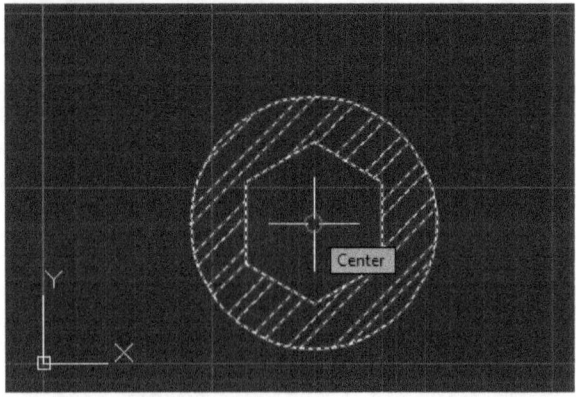

Pic 2.130 Selecting base point

4. Show the new position, you can use polar or relative coordinate.

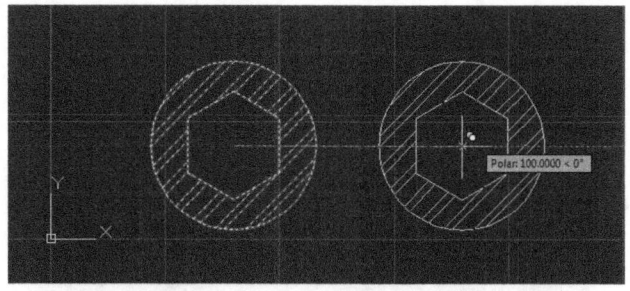

Pic 2.131 Specifying the new position

5. The copying result will be displayed in AutoCAD. And the initial object still exists.

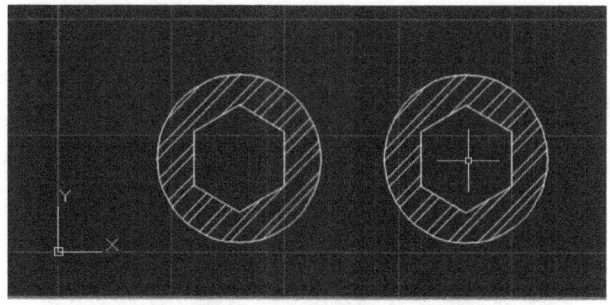

Pic 2.132 Copying result

2.2.7 Mirror

Mirror command mirrors object using a line as the mirror. See steps below for mirror command example:

1. Type "mirror' command.

2. Select object you want to mirror.

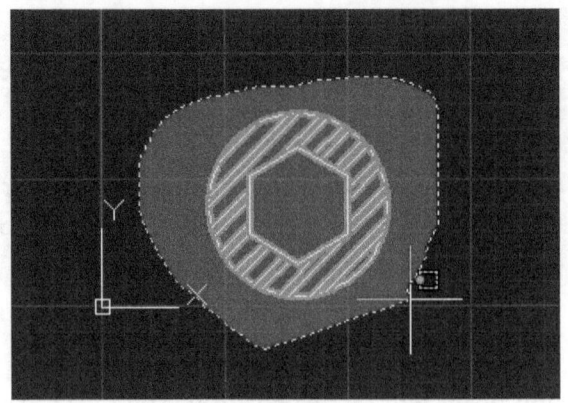

Pic 2.133 Selecting object

3. Selected object will become dotted line.

4. Specify the first line to make the mirror.

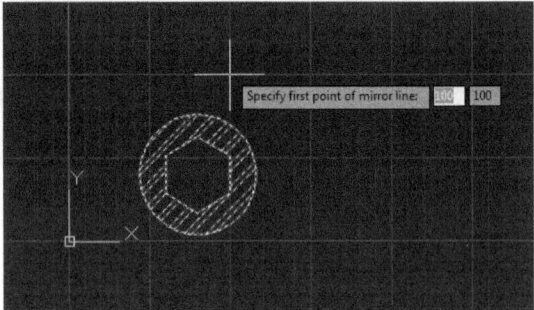

Pic 2.134 First line for the mirror

5. Specify the second line for the mirror.

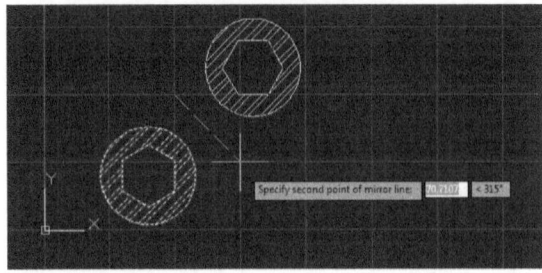

Pic 2.135 Specify the second line for the mirror

6. The object will be mirrored, and you'll be asked whether you want to initial object or not?

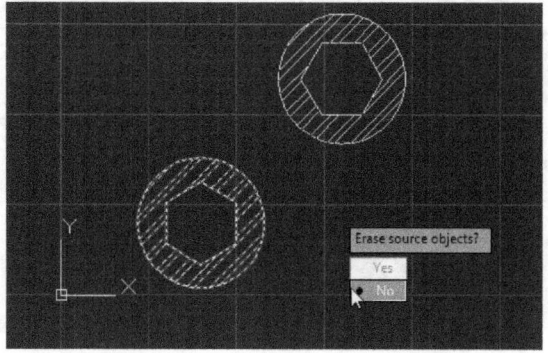

Pic 2.136 Option to erase initial object or not

7. You can see the initial object and mirrored object on the drawing area.

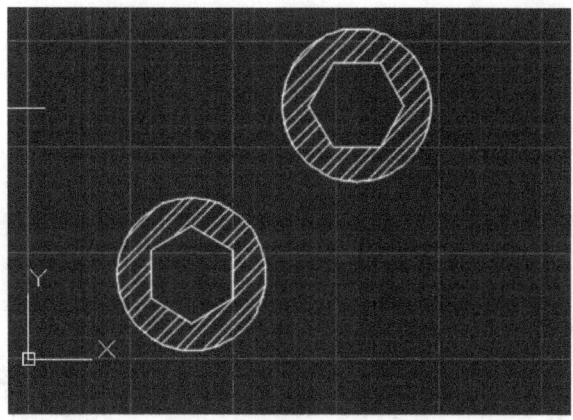

Pic 2.137 Mirroring area

2.2.8 Fillet

The fillet can be made from two line, see example below:

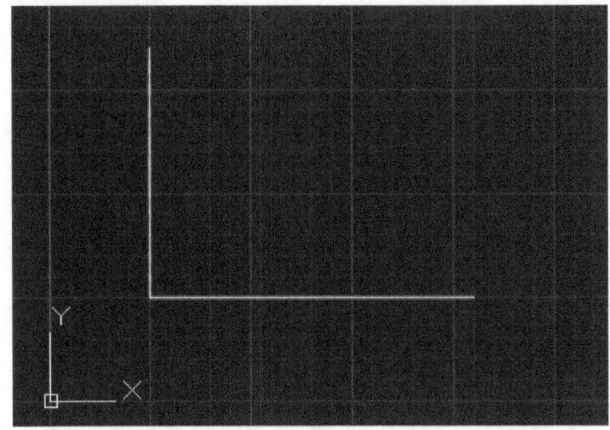

Pic 2.138 Line to be filleted

See example below on how to create fillet:

1. Run "fillet" command.

```
Command: FILLET
Current settings: Mode = TRIM, Radius = 0.0000
```

2. Click R and set fillet radius to 40.

```
Select first object or [Undo/Polyline/Radius/Trim/Multiple]: R
Specify fillet radius <40.0000>: 40
```

3. Click on the first line.

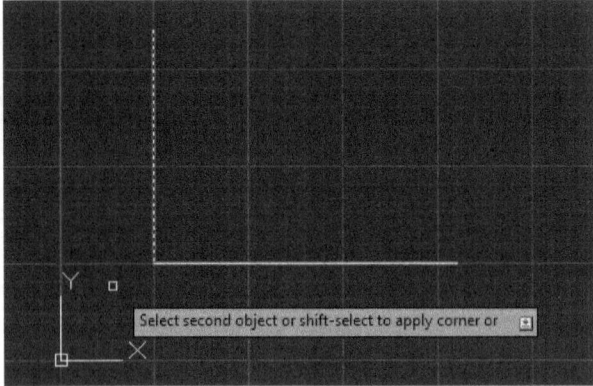

Pic 2.139 Click on the first line

4. Click on the second line.

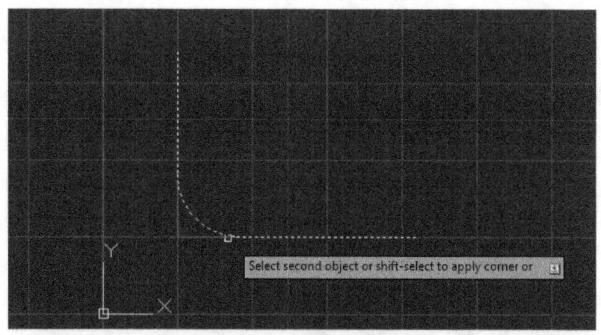

Pic 2.140 Second fillet

5. After you click the second line, fillet created automatically.

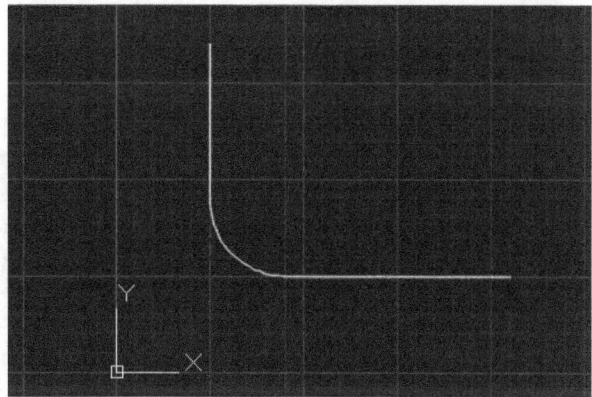

Pic 2.141 Fillet result

2.2.9 Chamfer

Chamfer similar to fillet, but chamfer is not an the arc, it's a line. See example below to create chamfer:

1. Execute "chamfer" command, click D to specify the distance of the chamfer.

```
Command: _chamfer
(TRIM mode) Current chamfer Dist1 = 0.0000, Dist2 = 0.0000
Select first line or [Un-
do/Polyline/Distance/Angle/Trim/mEthod/Multiple]: D
```

2. Set first distance to 40, and second distance to 40.

```
Specify first chamfer distance <0.0000>: 40
Specify second chamfer distance <40.0000>: 40
```

3. Click on the first line.

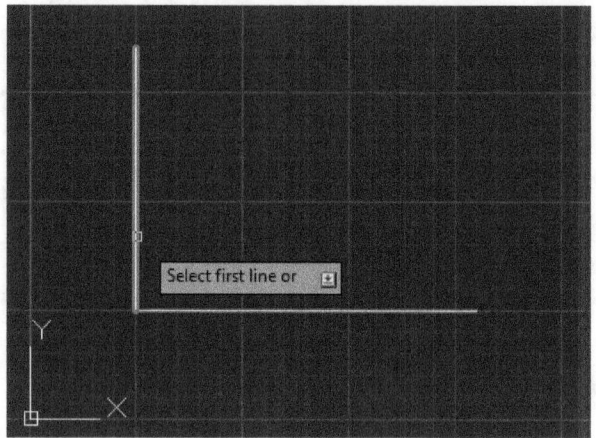

Pic 2.142 Click on the first line

4. Click on the second line.

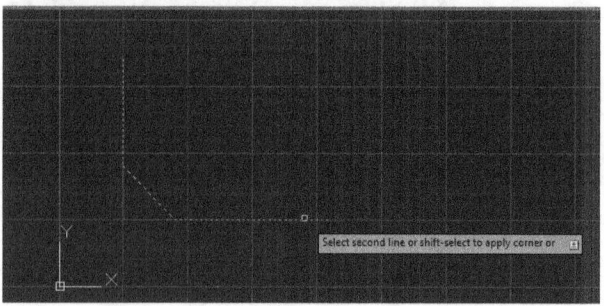

Pic 2.143 Click on the second line

5. See picture below for the chamfer result.

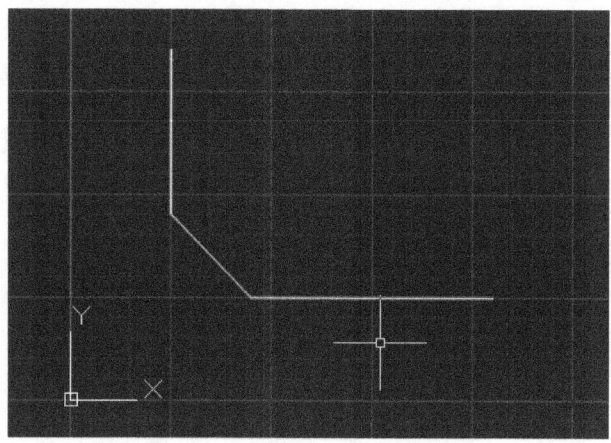

Pic 2.144 Chamfer result

2.2.10 Explode

Explode command explodes polyline or region to segments. See steps below:

1. There's a polyline:

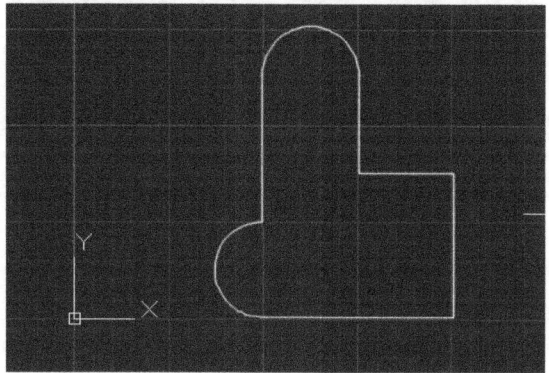

Pic 2.145 Polyline

2. If you choose a polyline, all segments will become a dotted line, this is because it's one object.

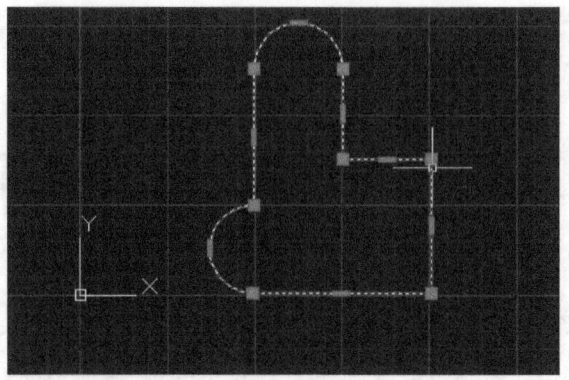

Pic 2.146 All segments of polyline become dotted line

3. Now execute "explode" function.

4. Select the polyline object.

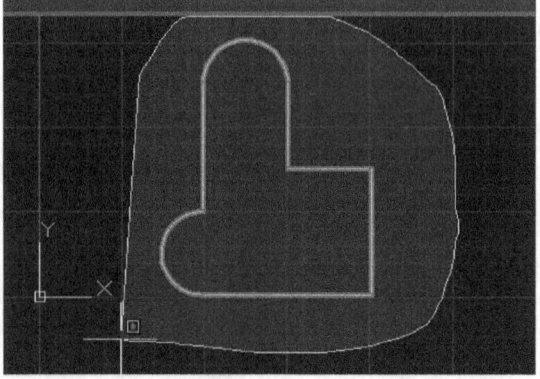

Pic 2.147 Polyline selection

5. Click Enter, the object will be exploded. If you click on the object, a segment will be selected. This means the object already segmented/exploded.

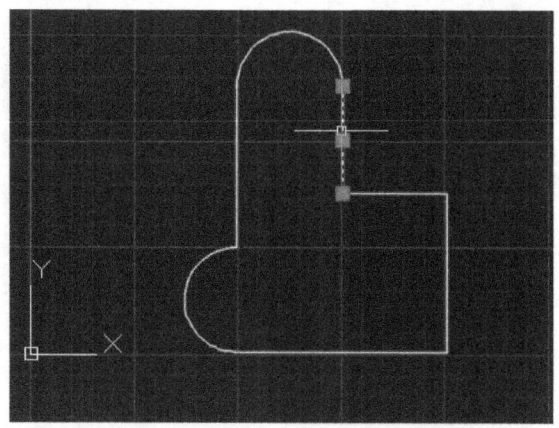

Pic 2.148 The object after segmentation

6. If you want to choose more than one segment, you have to click those segments, one by one.

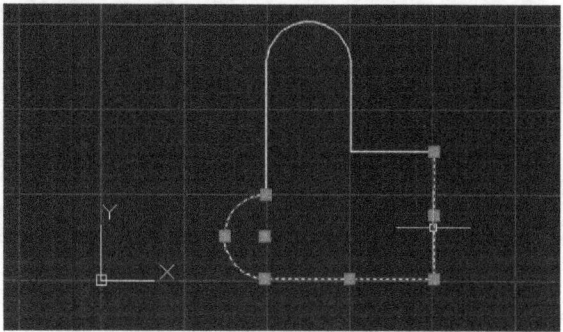

Pic 2.149 Choosing 3 segments after exploded

2.2.11 Stretch

Stretch command stretches object. You just have to define which object to stretch, see example below:

1. For example, there is an object as below:

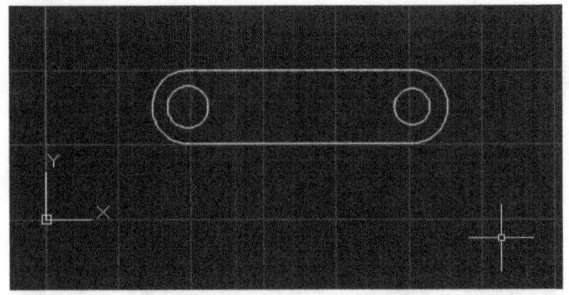

Pic 2.150 Object to stretched

2. Execute "stretch" command, and select part of the object you want to stretch.

```
Command: STRETCH
Select objects to stretch by crossing-window or crossing-polygon...
```

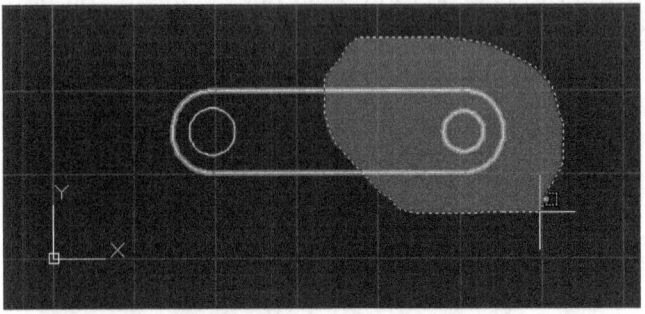

Pic 2.151 Choosing object to stretch

3. Selected object will be a dotted line.

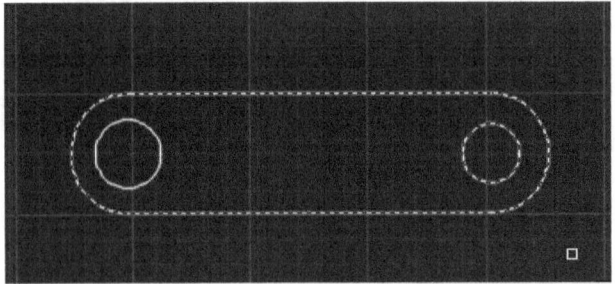

Pic 2.152 Selected object becoming dotted line

4. Click Enter, and specify the base point.

```
Specify base point
```

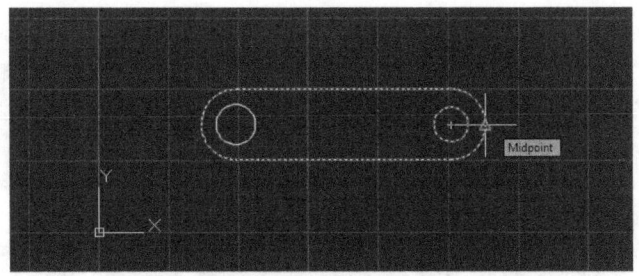

Pic 2.153 Specify base point for stretching

5. Click on a base point and click the second point.

```
Specify base point or [Displacement] <Displacement>:
Specify second point or <use the first point as displacement>:
```

6. Drag to right, you can see the initial position and position after stretching.

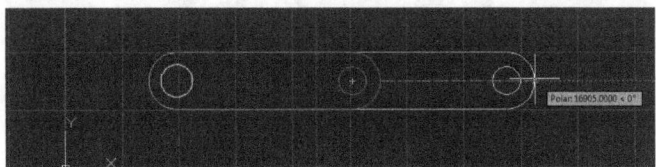

Pic 2.154 Stretching to right

7. If the mouse drag released, the object will be stretched.

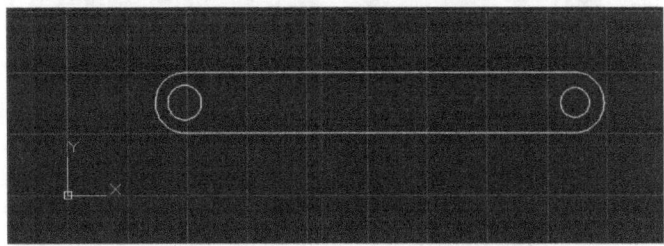

Pic 2.155 The object after stretching

8. Stretch can also be used for makes size smaller. By dragging to the left.

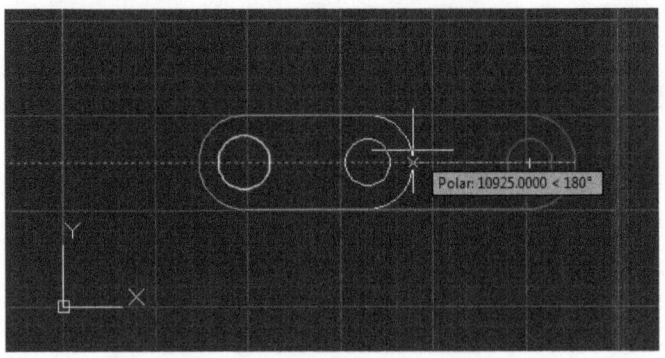

Pic 2.156 Negative stretching

9. If you do negative stretching, the object will be smaller.

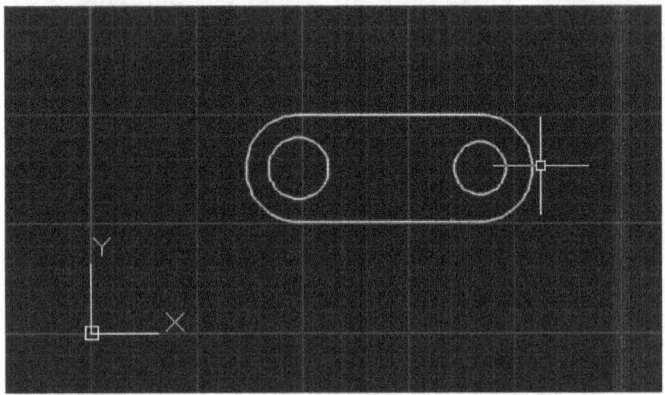

Pic 2.157 Negative stretch makes the object smaller

2.2.12 Scale

Scale command will scales object to make the object larger or smaller. See example below:

1. Type "scale".

2. Select the object.

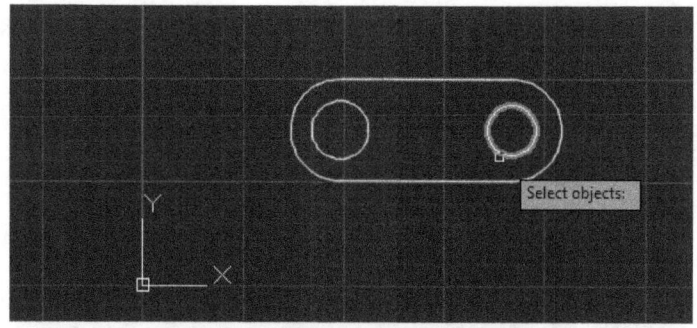

Pic 2.158 Choose the object to scale

3. Click Enter, the object will be a dotted line.

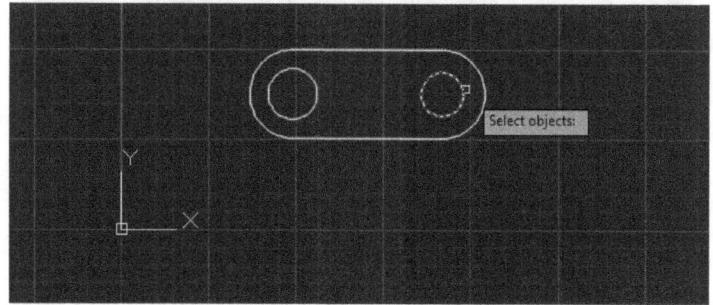

Pic 2.159 Object selected

4. Specify base point for scaling.

```
Specify base point
```

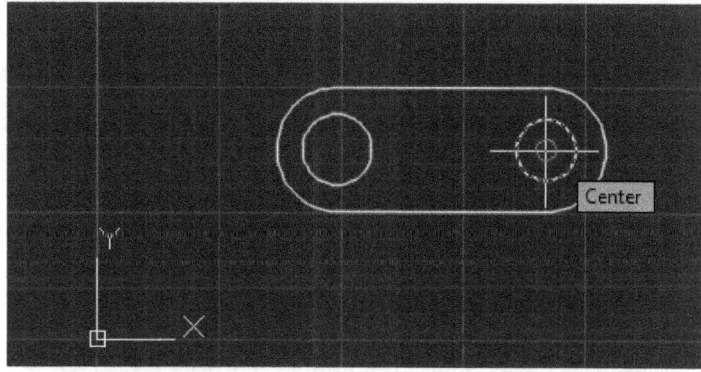

Pic 2.160 Click on the center as base point for scaling

---

5. Specify the scale factor or zoom factor, for example, if I take 2, it means the object will be zoomed twice.

```
Specify scale factor or [Copy/Reference]: 2
```

6. The result is:

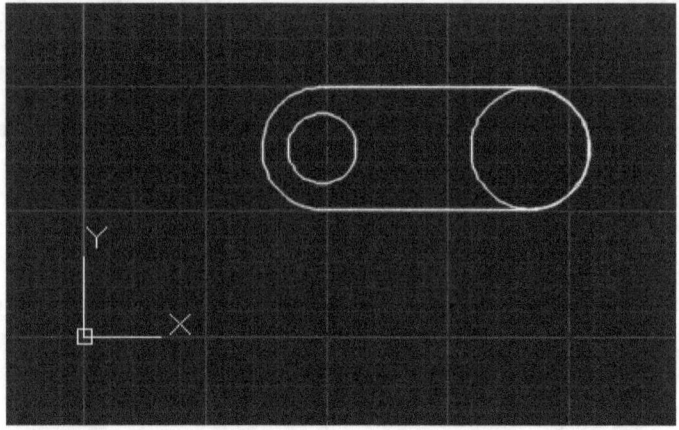

Pic 2.161 The result of scale factor

2.2.13 Array Rect

You can copy object and paste it in array of rows and columns by using array rect. This is how to use Array Rect command:

1. Type "arrayrect" in the command prompt.

2. Select the object.

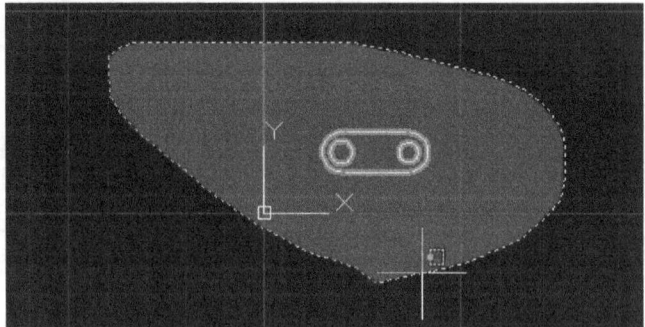

Pic 2.162 Select object you want to copy with array rect

3. Object automatically copied with array rect.

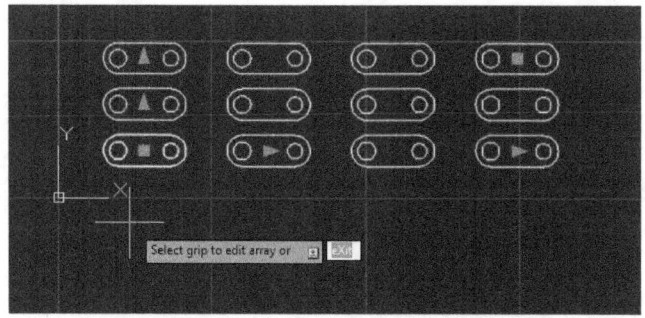

Pic 2.163 Object copied as array

4. You can change the property of array rect using column and row in **Columns** and **Rows**.

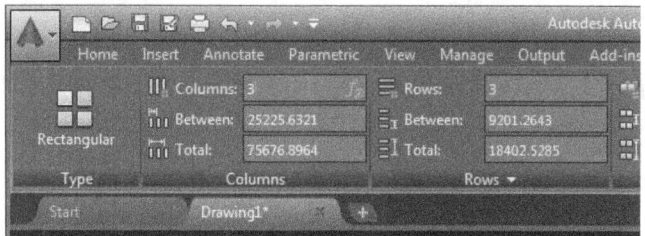

Pic 2.164 Columns and rows

5. Click **Close Array** to close the array creation.

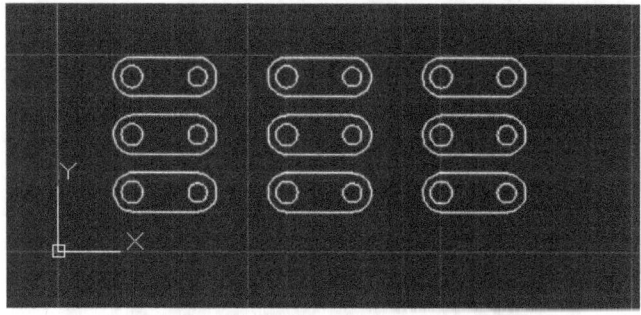

Pic 2.165 Array rect result

CHAPTER 3 CASE STUDIES

On this chapter, I'll demonstrate how to implement skills you have learned from the previous chapter to draw a simple drawing.

3.1   Create Simple House Plan

For example, you will create simple house plan with size 100x100. See steps below:

1. Set limits from workspace from 0,0 to 100, 100.

```
Command: LIMITS
Reset Model space limits:
Specify lower left corner or [ON/OFF] <0.0000,0.0000>: 0,0
Specify upper right corner <100.0000,100.0000>: 100,100
```

2. Draw a line as below:

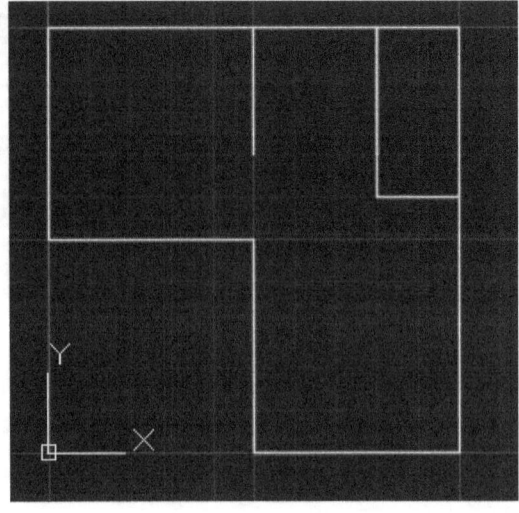

3. See pic above, the size is 100 x 100.

4. Create small rectangle with size = 2.5 x 2.5.

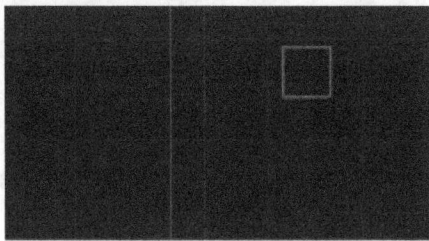

Pic 3.2 Small rectangle

5. Type Move, and click the object.

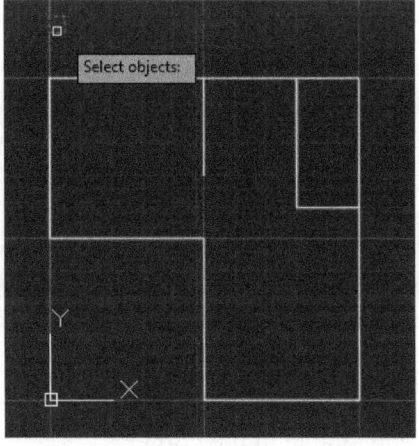

Pic 3.3 Small object

6. Choose the midpoint of the little rectangle as a base point.

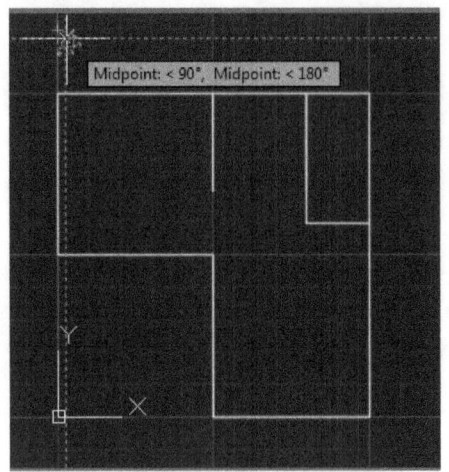

Pic 3.4 Choosing base point of move

7. Put the small rectangle to the corner of each line.

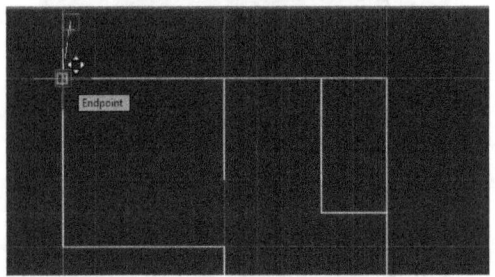

Pic 3.5 Put small rectangle to the corner

8. The box will be available in the corner.

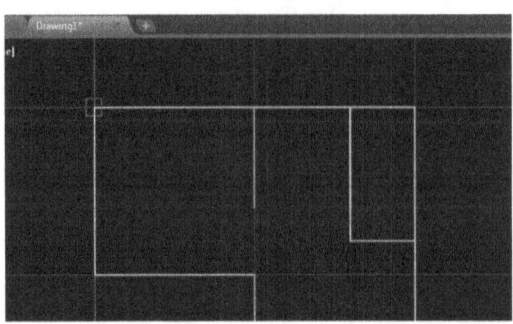

Pic 3.6 Box in the corner

9. Type "Copy" and select the small object.

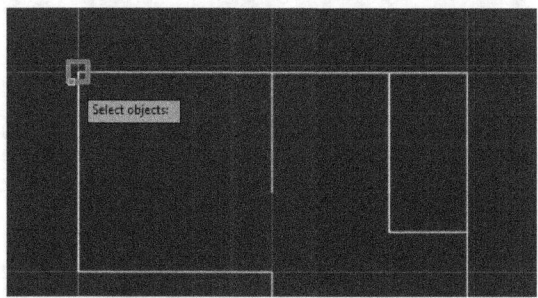

Pic 3.7 Choose the small rectangle to copy

10. The small rectangle will become dotted line.

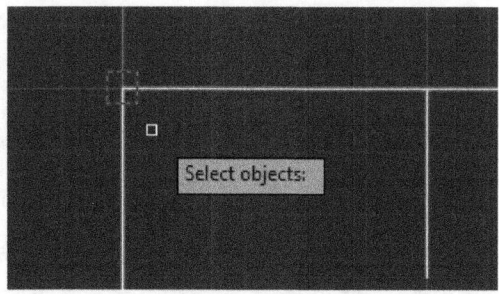

Pic 3.8 Small rectangle selected

11. Click on the mid rectangle when you are asked: **Specify base point**.

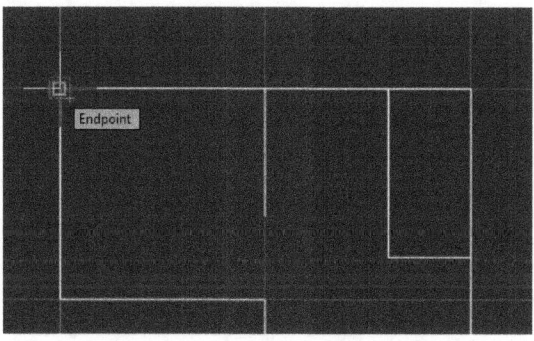

Pic 3.9 Specify base point for copying

12. Then choose other corner/intersection point in **Specify end point**,

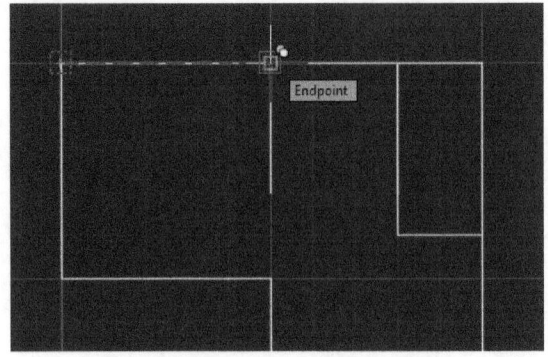

Pic 3.10 Specify the end point

13. Do this in each intersection/corner.

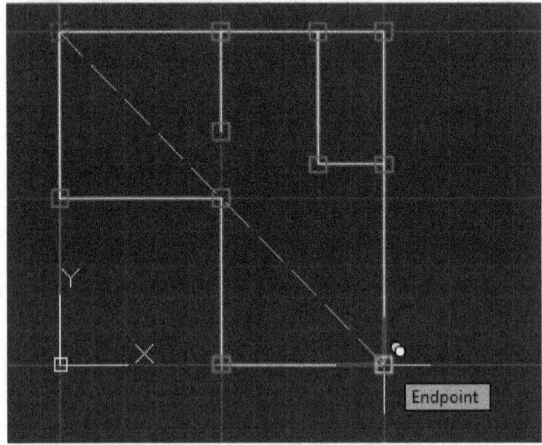

Pic 3.11 Copying the small rectangle in all corner/intersection

14. The result will be like the picture below:

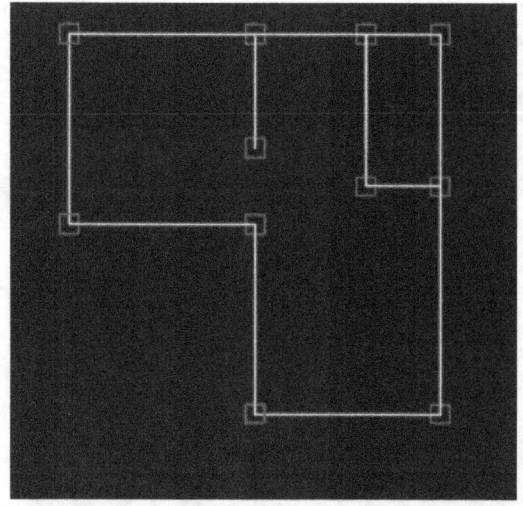

Pic 3.12 Small rectangle copied to each corner

15.   Draw a line to draw the wall.

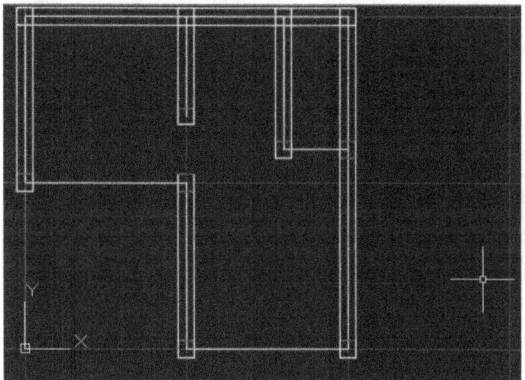

Pic 3.13 Drawing the wall

16.   You can also draw a line to make a border.

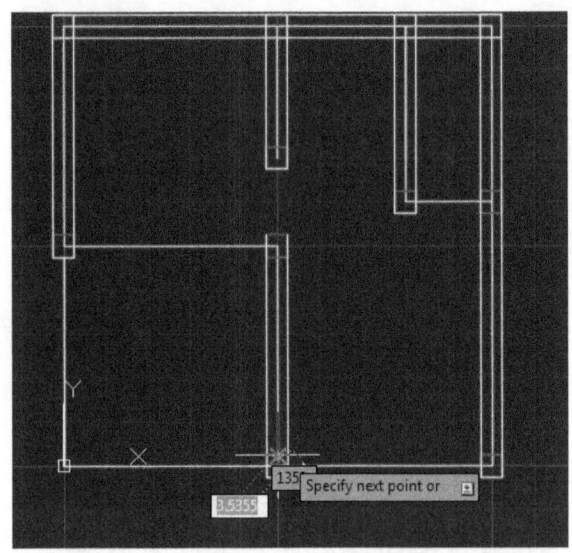

Pic 3.14 Draw a line to draw border and wall

17. Draw a door like this.

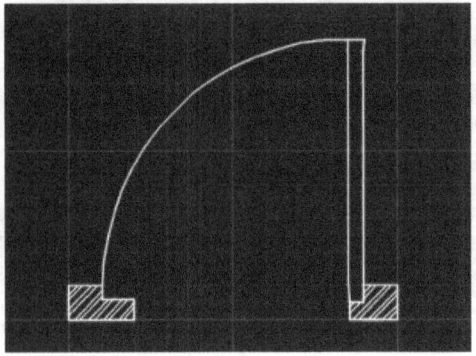

Pic 3.15 Draw a door

18. Move the door to the place you want to create door.

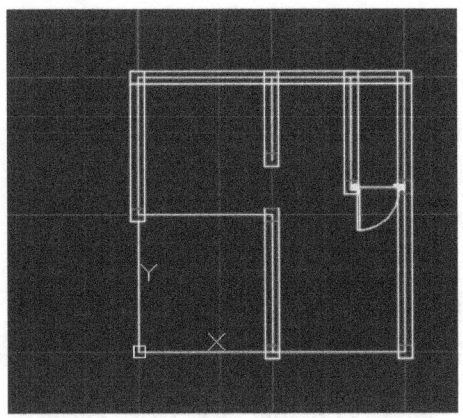

Pic 3.16 Create door

19. To give grass effect, create hatch, and select the pattern to Grass.

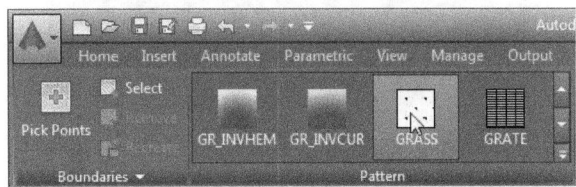

Pic 3.17 Choose pattern to grass

20. Give grass to the area you want to draw a grass.

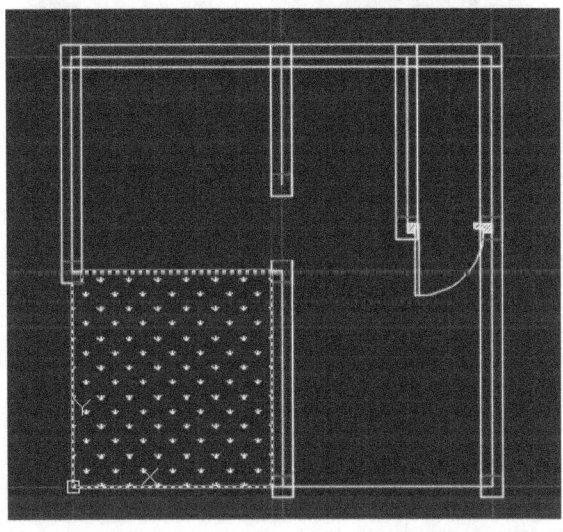

21. To give a car, click on **View > Tool Palettes**.

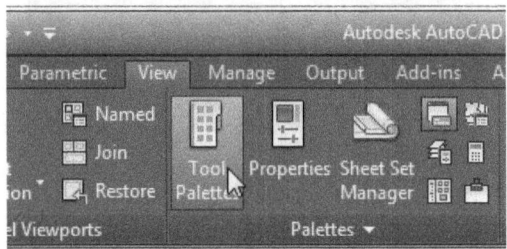

Pic 3.19 Click on View > Tool Palettes

22. Choose **The architectural > Vehicles**. Right click and select **Properties**.

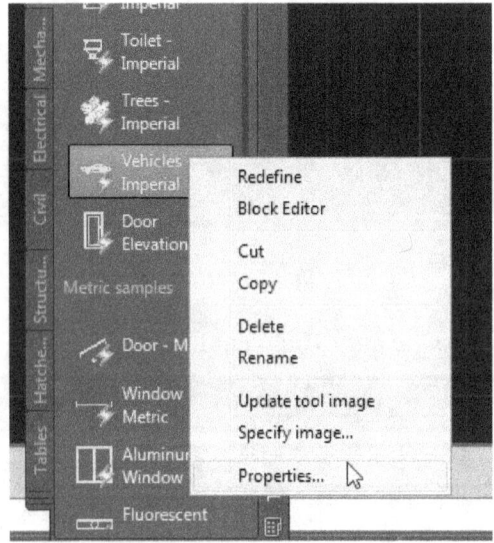

Pic 3.20 Click Properties

23. Choose **Type (view) to Sports Car (Top)**.

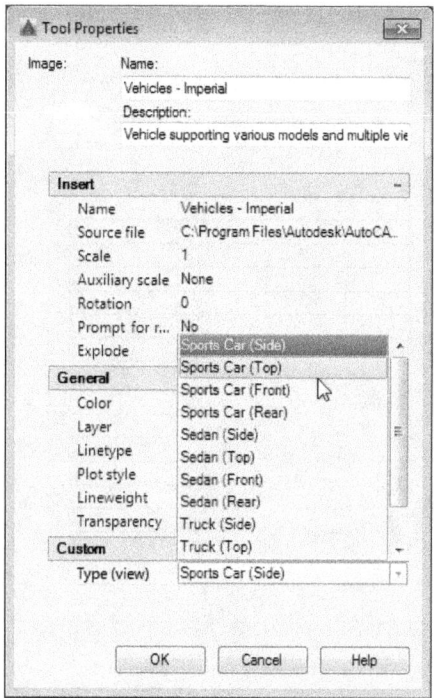

Pic 3.21 Choosing Sports Car (Top)

24. You can see the Type (View) changed and click OK

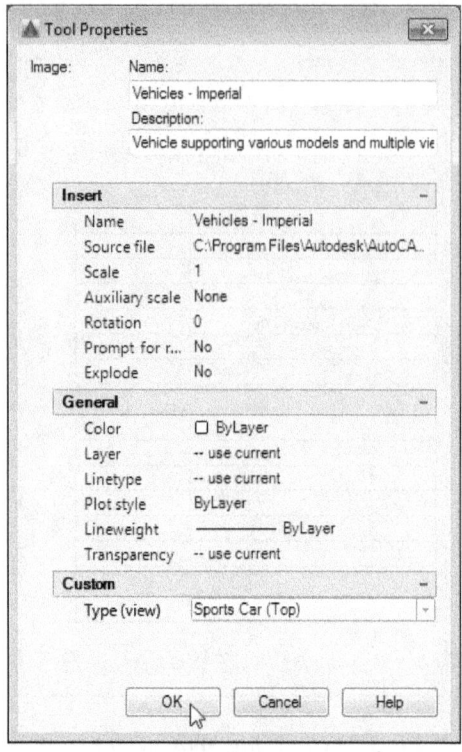

Pic 3.22 Type (view) property for car object already changed

25.   Click to insert car object.

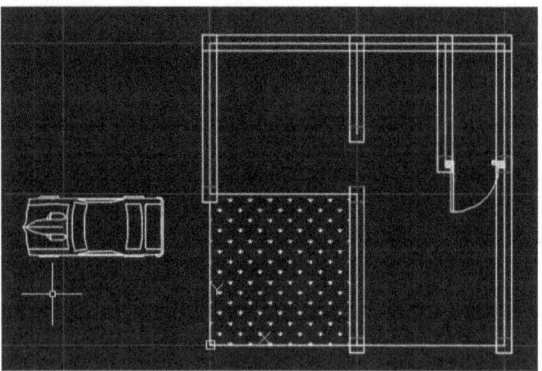

Pic 3.23 Car object inserted to drawing

26.   Rotate using rotate function and put it in the garage.

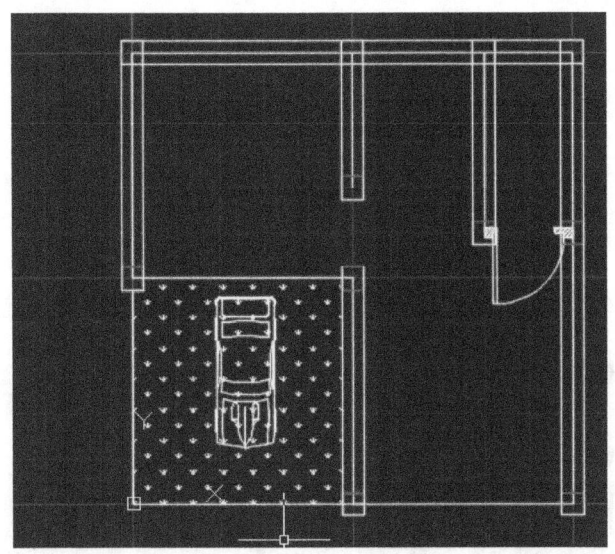

Pic 3.24 Put the car object

27. Using the same method, you can add other objects, like tree.

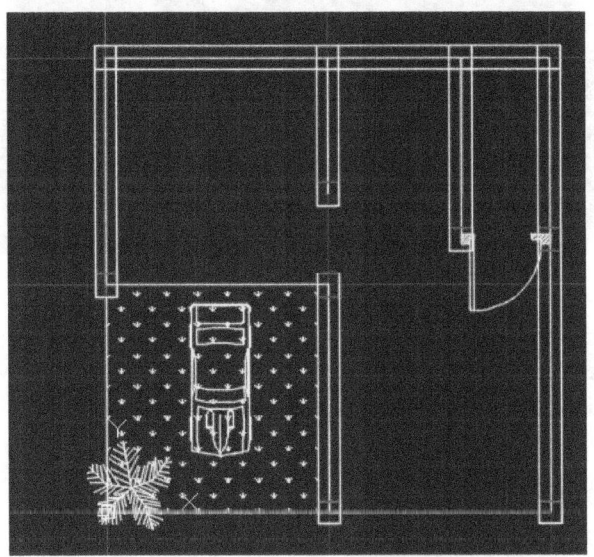

Pic 3.25 Inserting another object

28. To insert annotations, click **Home > Annotation**.

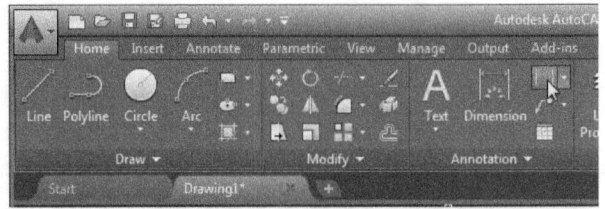

Pic 3.26 Annotation box

29. Complete the annotation in another place.

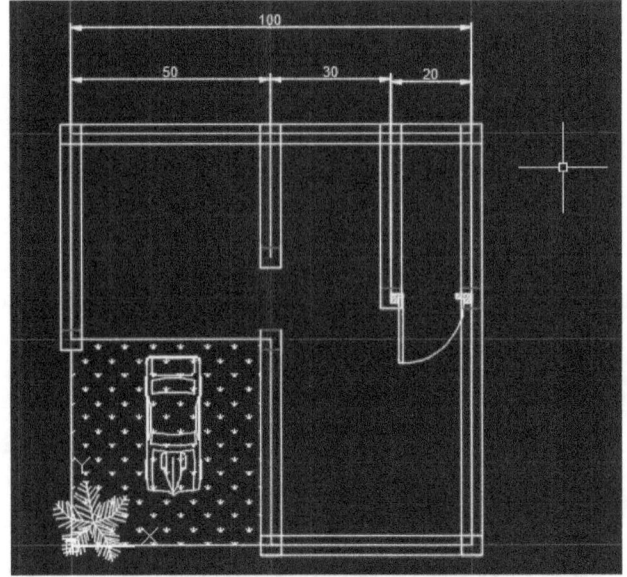

Pic 3.27 Completing the annotation

30. You can create other objects to complete the drawing using polyline, circle and rectangle.

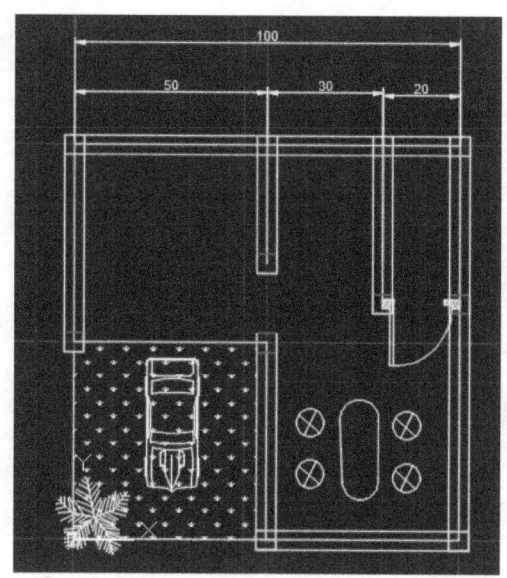

Pic 3.28 Completing the object

31.   You can again add more annotation.

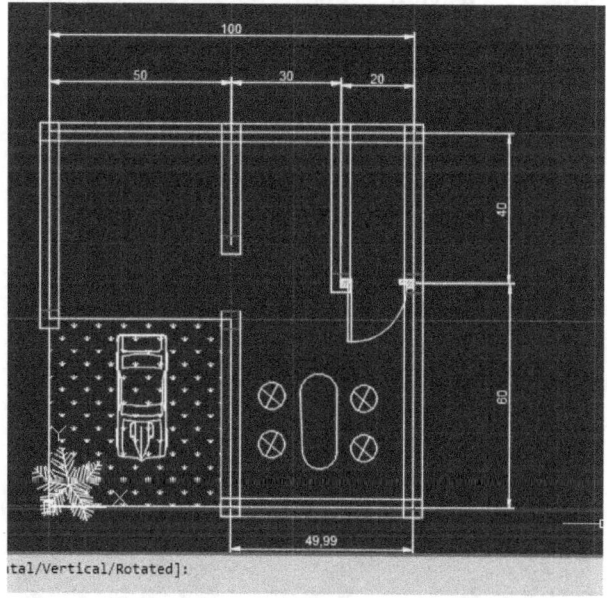

Pic 3.29 Adding annotations on other places

32.   The result will be like this, you can add more using your creativity.

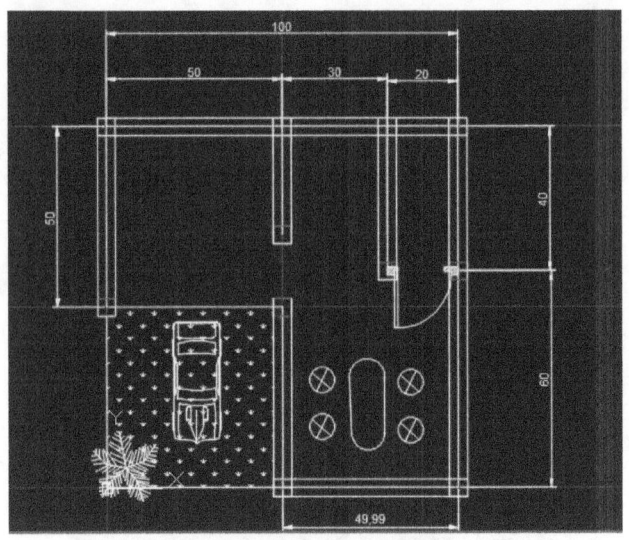

Pic 3.30 Final result of house plan drawing

3.2   Create Simple Gear

In this tutorial, you'll learn about how to create simple gear, follow steps below:

1. Create two circles, with an identical center point, but with different radius. Then add the teeth of the gear.

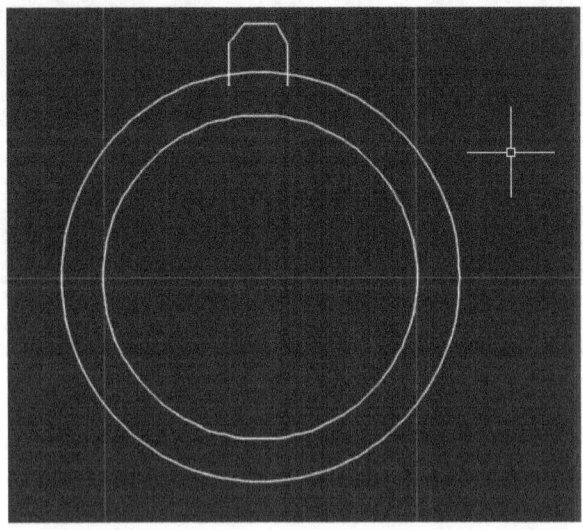

Pic 3.31 Create two circle with identical center point

2. Trim the root of the teeth by entering Trim command then select all objects.

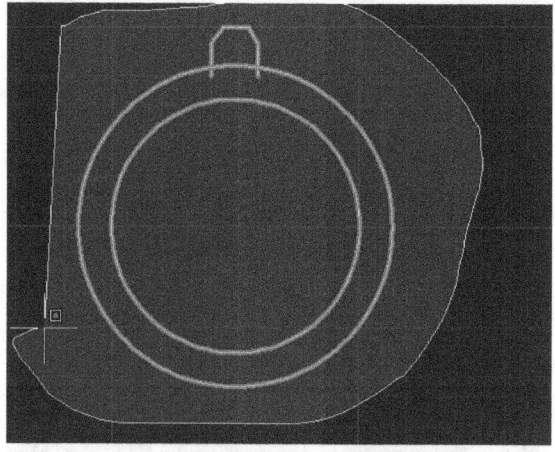

Pic 3.32 Selecting all objects to trim

3. Click on the root of the teeth to trim it.

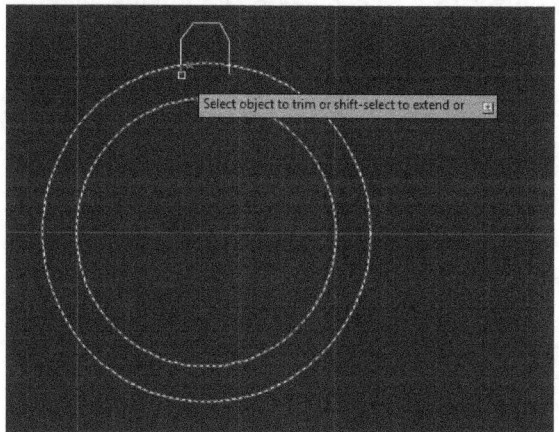

Pic 3.33 Trim the tooth's root

4. You can see the tooth.

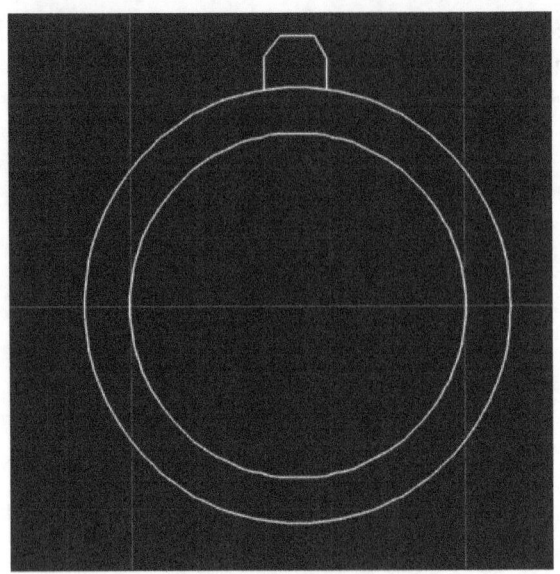

Pic 3.34 Circle with one tooth

5. Copy the object 18X, execute copy command, and select objects.

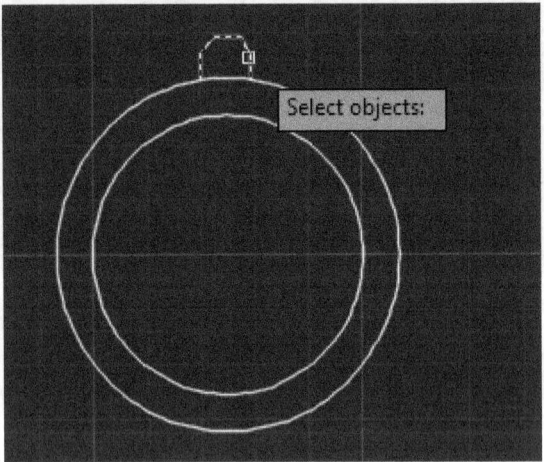

Pic 3.35 Selecting object to copy

6. Then rotate the copied tooth using Rotate command, choose the object then specify a base point to the center.

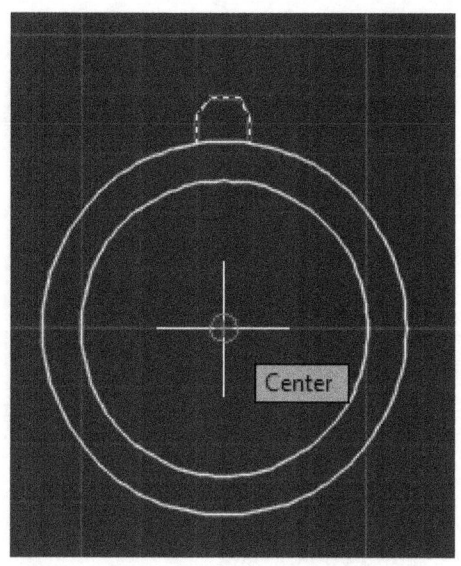

Pic 3.36 Specify base point = the center

7. Rotate with 20 degrees interval.

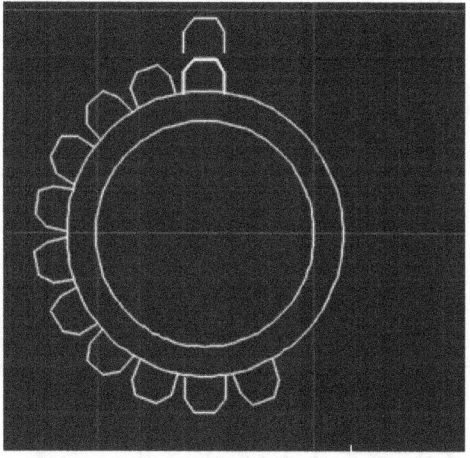

Pic 3.37 Rotating the teeth with 20 degrees' interval

8. Do until all the teeth rounding the circle.

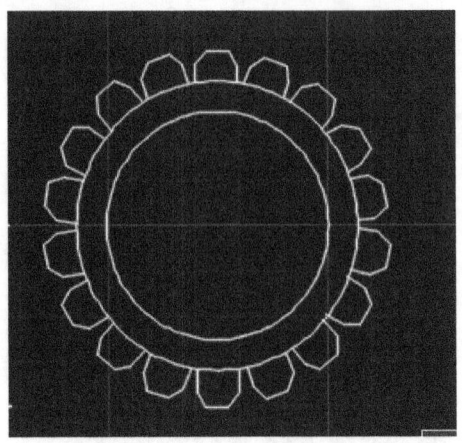

Pic 3.38 Teeth rounding the circle

9. Type "linetype", and click Load.

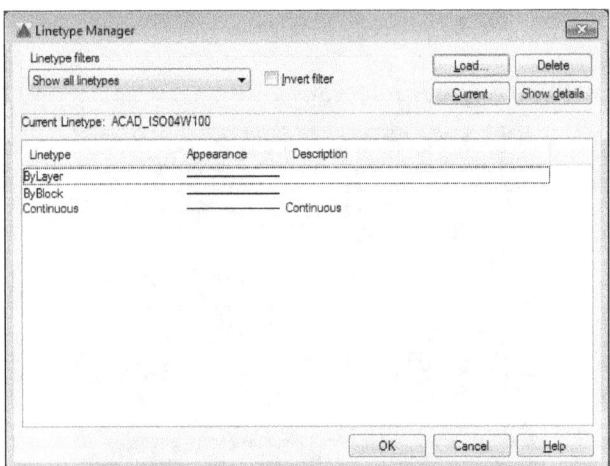

Pic 3.39 Choosing the linetype

10. Choose ISO long-dash dot to draw the axis.

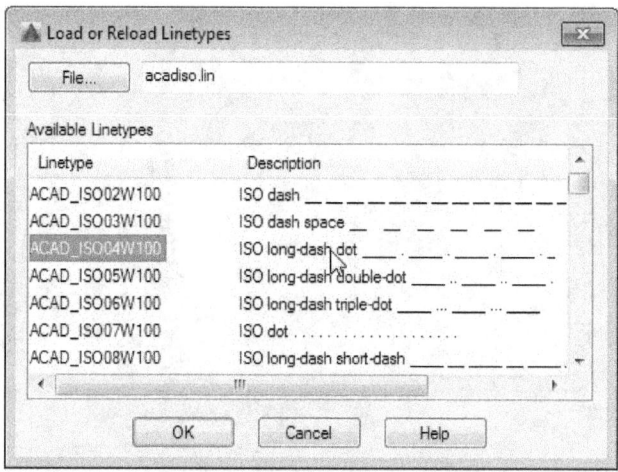

Pic 3.40 Adding ISO long-dash dot

11.   Then click on the iso long-dash dot, and click Load.

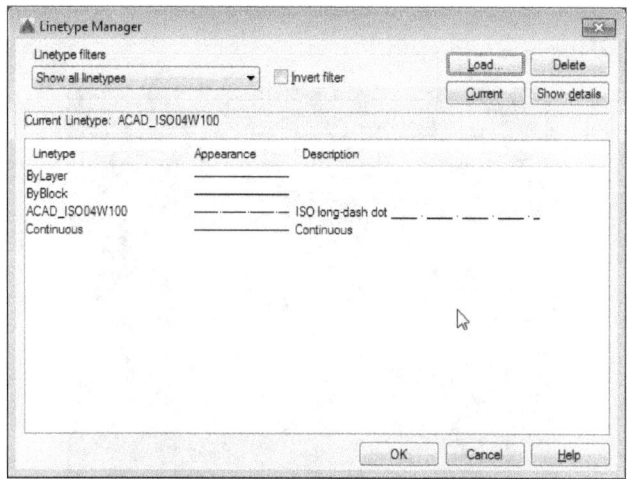

Pic 3.41 Choosing Longtype dash dot

12.   Draw vertical axis.

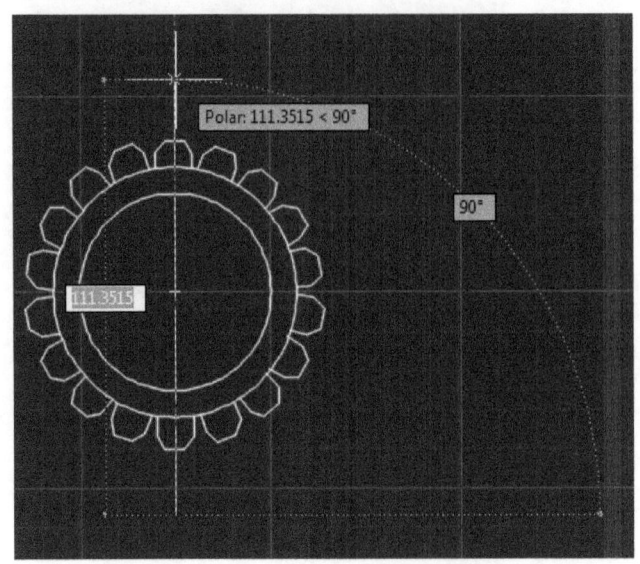

Pic 3.42 Drawing vertical axis

13. Draw horizontal axis.

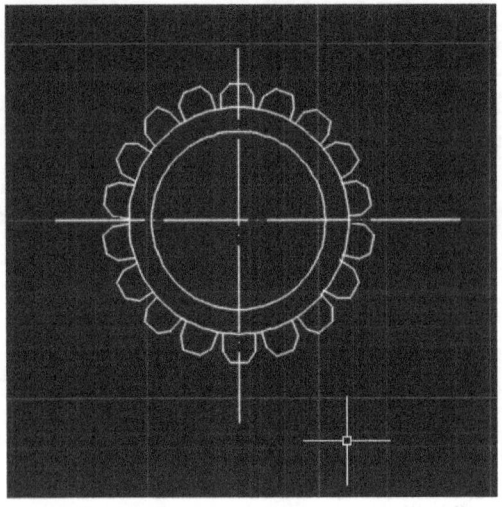

Pic 3.43 Drawing horizontal axis

14. Trim the gear, and select all the objects.

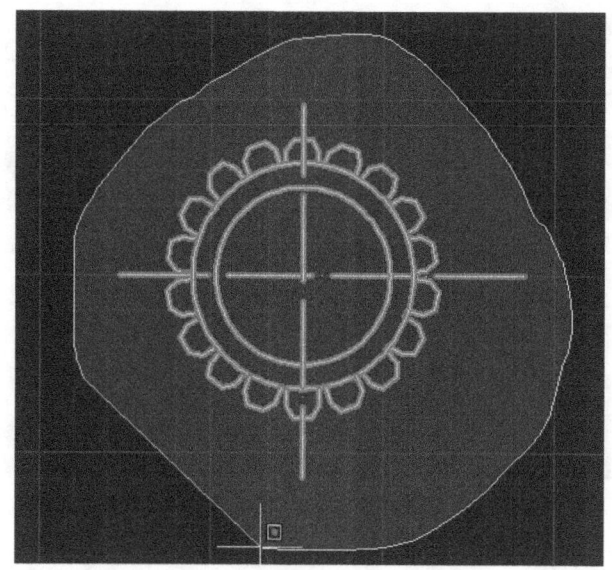

Pic 3.44 Select the gear

15. Click on the line below the teeth.

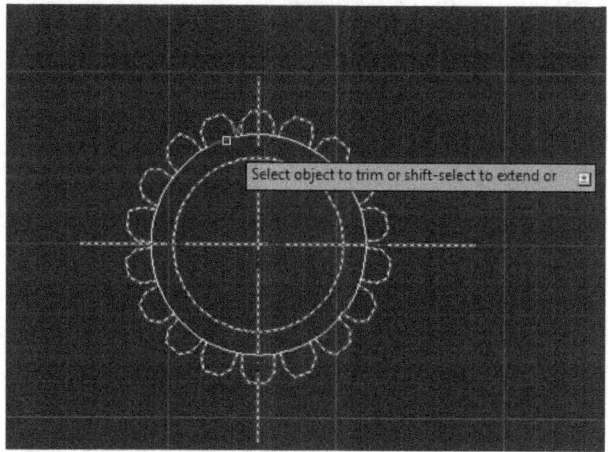

Select object to trim or shift-select to extend or

Pic 3.45 Trimming the line below the teeth

16. The final result will be as below:

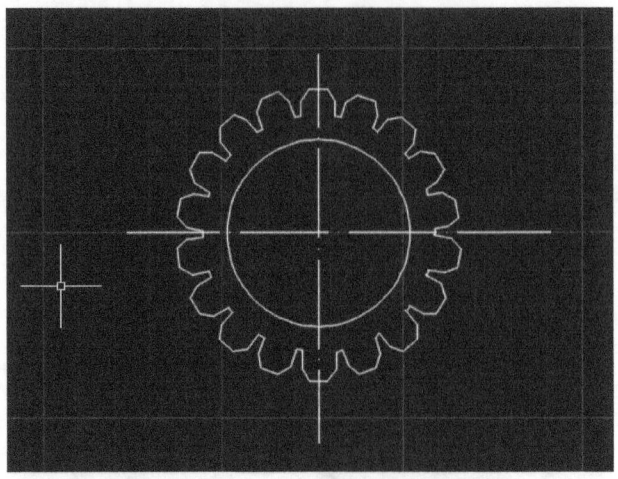

Pic 3.46 Final result creating gear

3.3   Create Simple Piston

See example below for creating simple piston using AutoCAD:

1. Create two circles, and two lines.

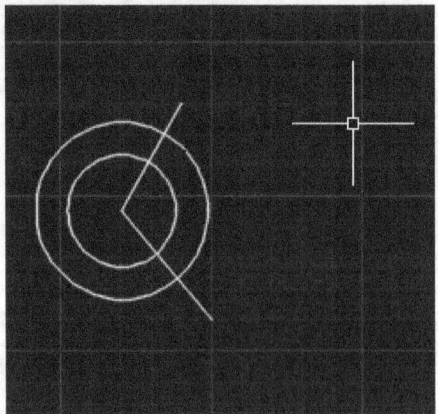

Pic 3.47 Create two circles and two lines

2. Type "Trim" and select all objects.

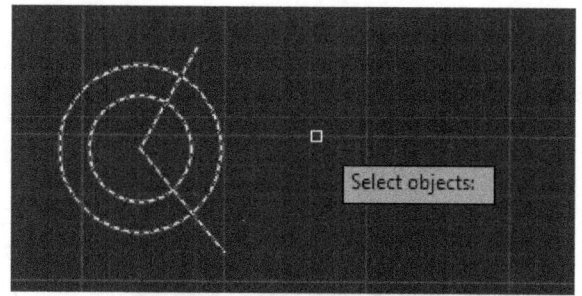

Pic 3.48 Choosing all objects to trim

3. Trim to make picture below:

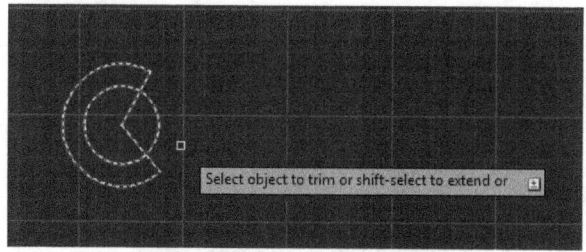

Pic 3.49 Trim outer circle

4. Trim part of the inner circle, see picture below:

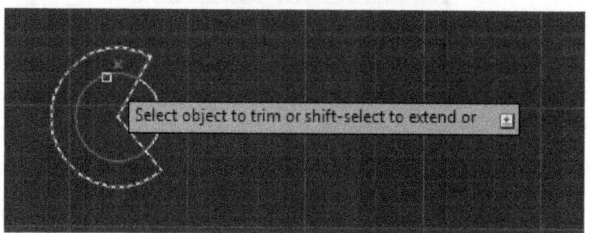

Pic 3.50 Trim inner circle

5. Trim the radius line.

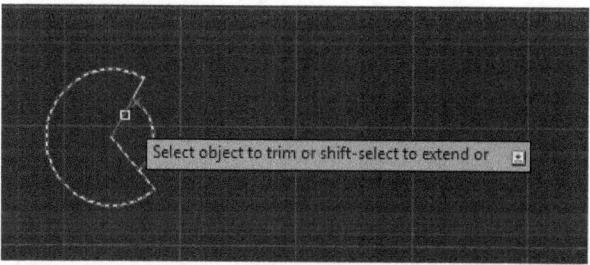

6. The result will be like this.

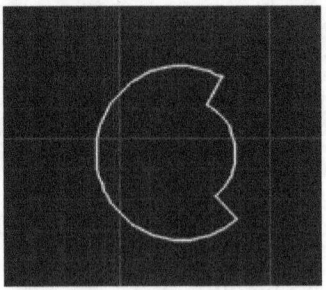

Pic 3.52 Engine axle drawing

7. Draw a small circle, with center point identical with the center point of the axle.

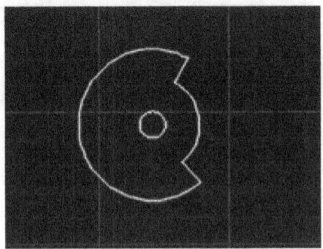

Pic 3.53 Small circle

8. Create one more small circle.

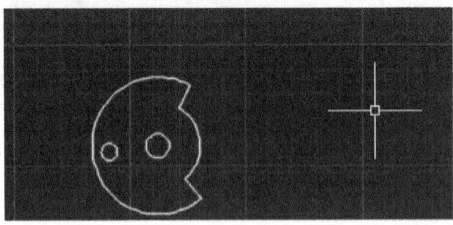

Pic 3.54 Creating one more small circle

9. Create polyline like picture below:

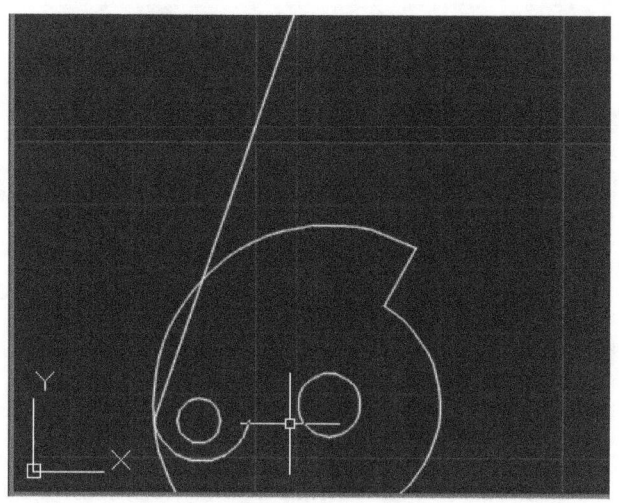

Pic 3.55 Create polyline

10.   Add the polyline with line and the arc.

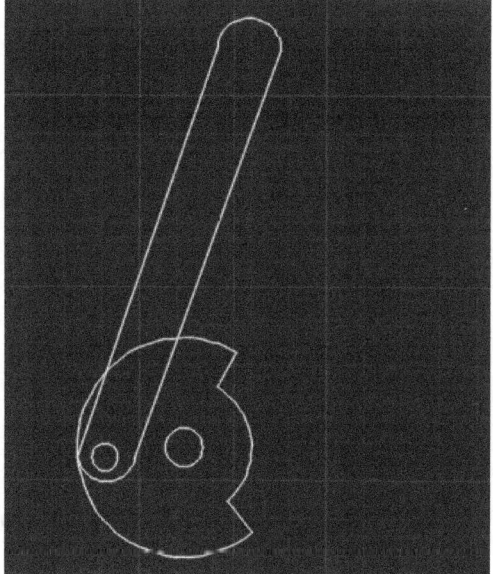

Pic 3.56 Create polyline with line and the arc

11.Create polyline to create the piston.

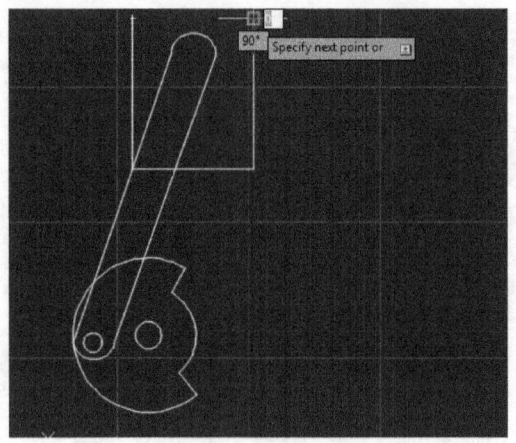

Pic 3.57 Create polyline to draw the piston

12. Draw an the arc to form the top of the piston.

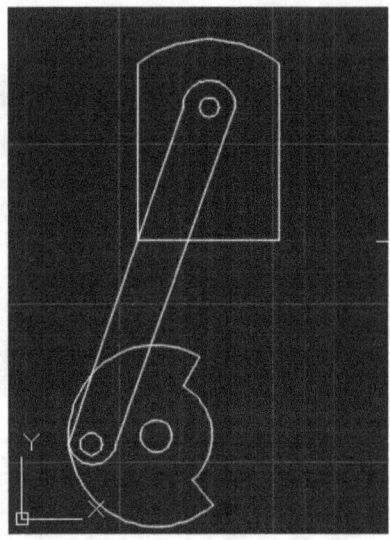

Pic 3.58 Draw an arc to form the top of piston

13. Now use trim.

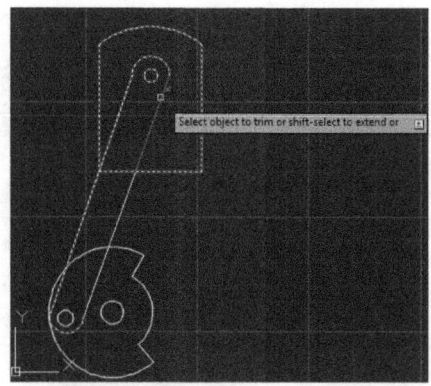

Pic 3.59 Trimming

14. The result after trim look as following picture.

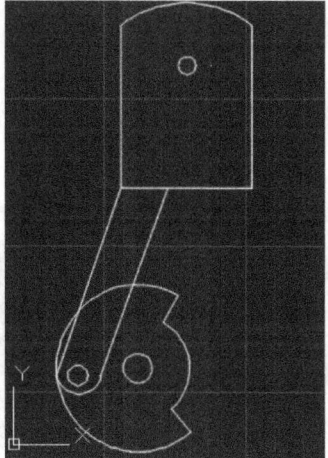

Pic 3.60 Result after trimming

15. Create two rectangles to draw the piston rings.

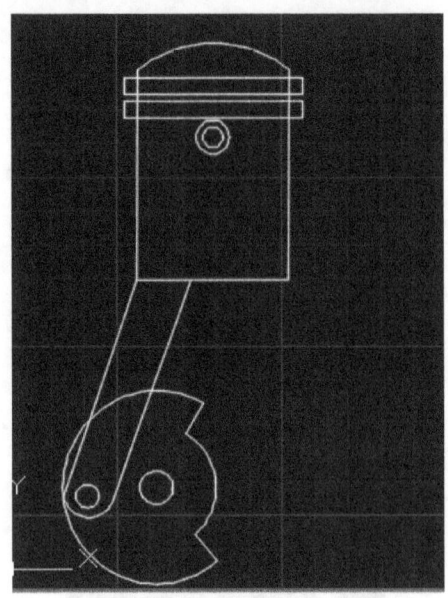

Pic 3.61 Create two rectangles

16.   Type trim, and select to trim.

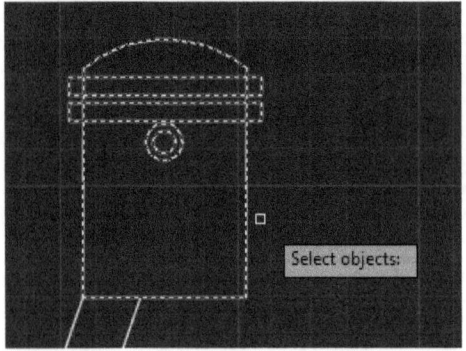

Pic 3.62 Select objects to trim

17.   Trim on the wall side of the piston inside the piston ring.

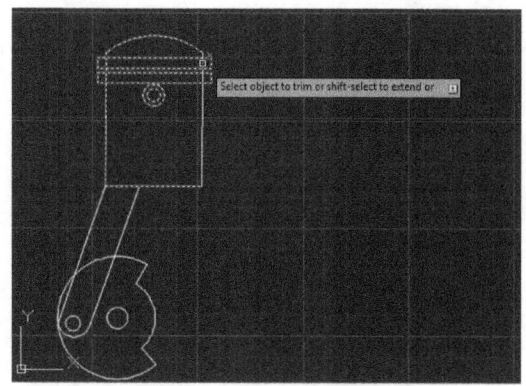

Pic 3.63 Trimming on piston ring

18. The final result is:

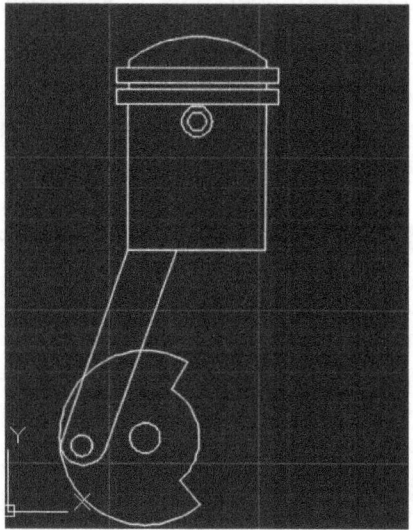

Pic 3.64 Final result

CHAPTER 4 DRAW 3D DRAWING

On this chapter, you'll learn how to create Pada bab ini, Anda belajar cara membuat 3D dasar dalam ruang kerja Modeling 3D. Anda dapat menggunakan bentuk tiga dimensi (3D) dari benda padat untuk membuat kotak, kerucut, silinder, bola, cincin, irisan, dan piramida.

Untuk membuat 3D yang solid, ubah workspace untuk 3D Modeling yang disesuaikan untuk membuat dan memodifikasi model 3D solid. Di akhir bab ini juga dijelaskna beberapa shortcut untuk bekerja lebih efektif dengan AutoCAD.

4.1   Configure 3D Workspace

Do steps below for configuring 3D workspace:

1. In status bar, click on Workpsace Switching.

Pic 4.1 Workspace Switching

2. In the menu, click 3D Basics

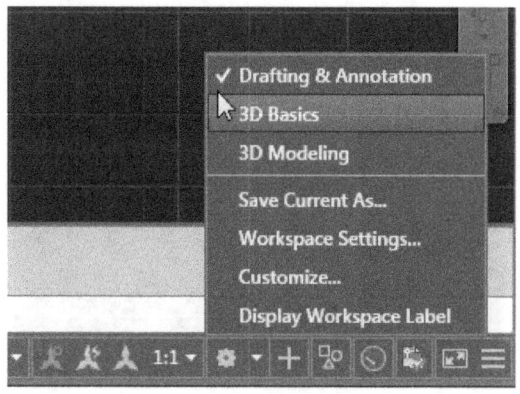

Pic 4.2 Clicking 3D Basics

3. 3D Basics workspace displayed, you can access lots of command and tools for drawing 3D objects.

4.2 Draw 3D Objects

Similar to 2D drawing, there are some basic objects in 3D Drawing. You'll learn how to draw 3D objects below:

4.2.1 Draw Box

Box is a rectangle with height. Here are steps to create box in atuocad:

1. Click Box icon on icon Create toolbar.

Pic 4.3 Click Box icon

2. Insert the first point, and insert the second point.

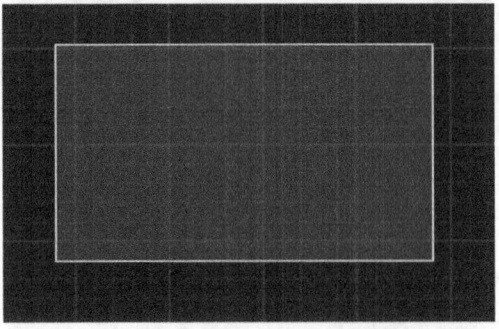

Pic 4.4 Create rectangle for box

3. Drag mouse to the top-right.

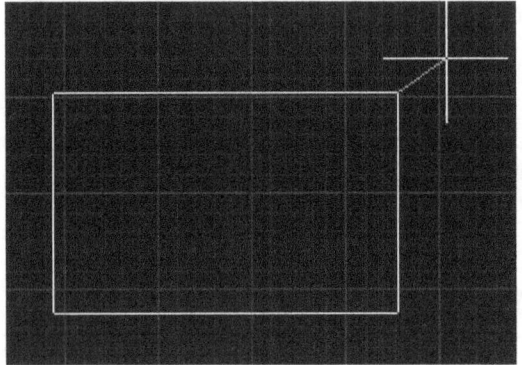

Pic 4.5 Drag mouse to top right

4. Insert the height of the box, for example 300.

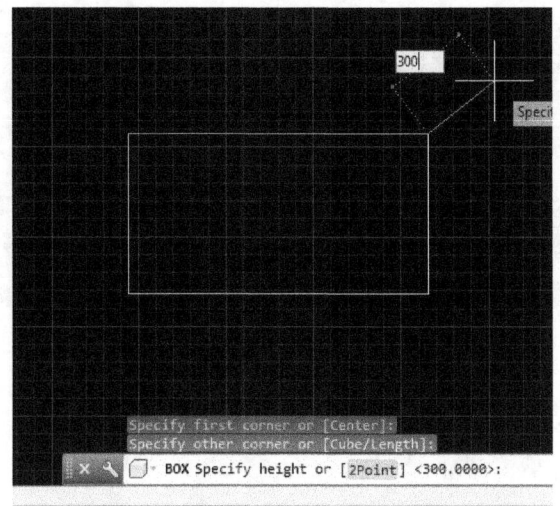

Pic 4.6 Specify the height of the box

5. To see the result in 3D, change the orbit button on the right toolbar.

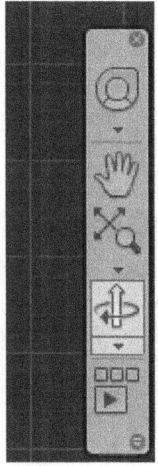

Pic 4.7 Changing the orbit

6. The result as below:

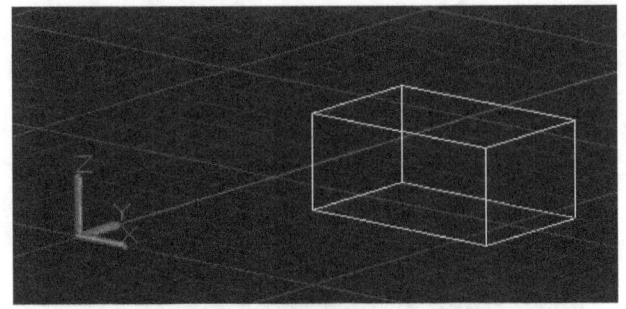

Pic 4.8 Result of box creation

7. Click Esc or [ENTER] in keyboard.

Other example:

1. Repeat step number 1-2.

2. In the command prompt, AutoCAD asks to specify other corder or length. Choose L for length.

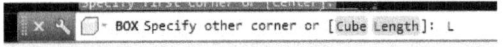

Pic 4.9 Choose L

3. It means, we'll draw by inserting length.

4. AutoCAD asks the length, type: 100.

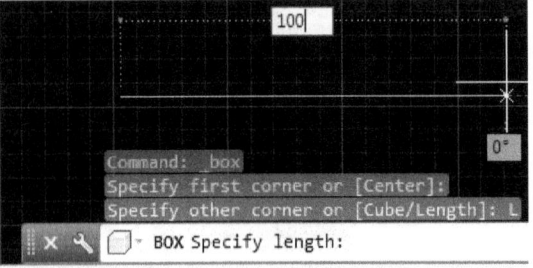

Pic 4.10 Displaying object's length

5. AutoCAD asks the width, type: 40.

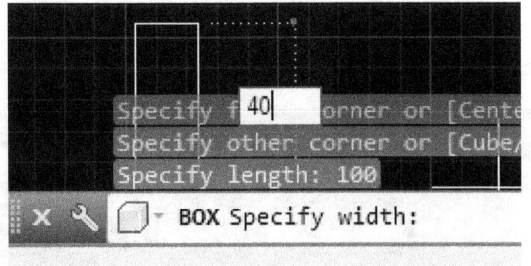

Pic 4.11 Specify width

6. Then drag top-right and specify the height to 50.

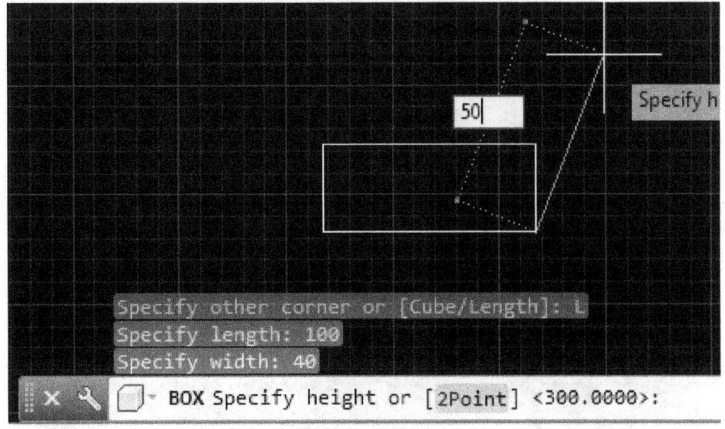

Pic 4.12 Specify height

7. The box result will be 100 x 40 x 50.

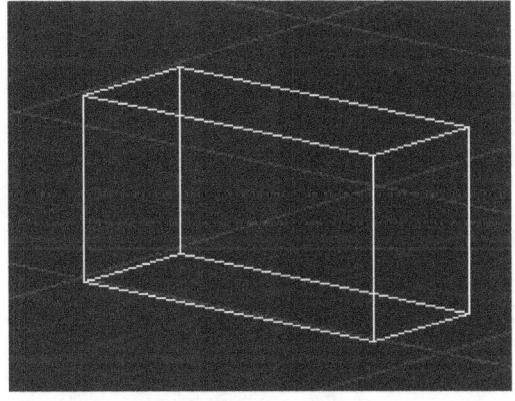

Pic 4.13 Box created

For drawing a cube, you can do these steps:

1. Execute Box, then type C for choosing cube.

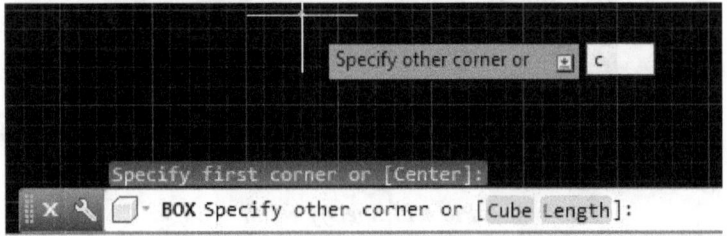

Pic 4.14 Specify C for cube

2. Specify the length to 100.

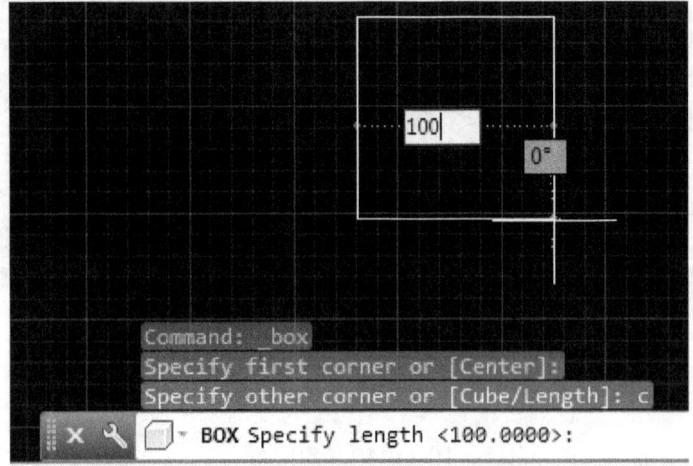

Pic 4.15 Specify the length of cube

3. The result is a cube with size: 100 x 100 x 100.

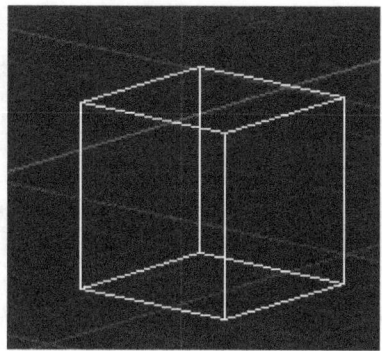

Pic 4.16 Cube already created

4.2.2 Draw Cylinder

The cylinder on AutoCAD created using Cylinder icon. See steps below:

1. Click on cylinder icon.

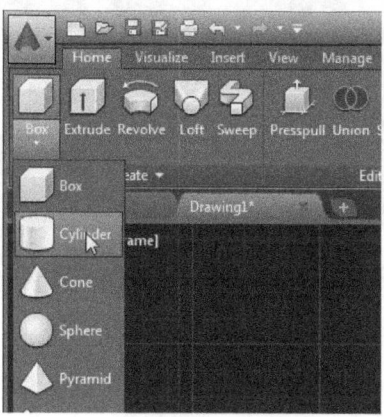

Pic 4.17 Click on Cylinder icon

2. Enter the center of cylinder. Click on a certain place.
3. Drag mouse to draw a circle.

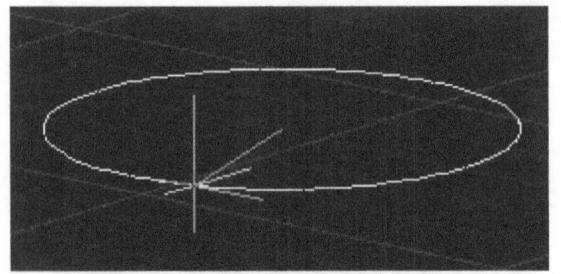

Pic 4.18 Drag mouse to draw a circle

4. Insert the height of the cylinder: 500.

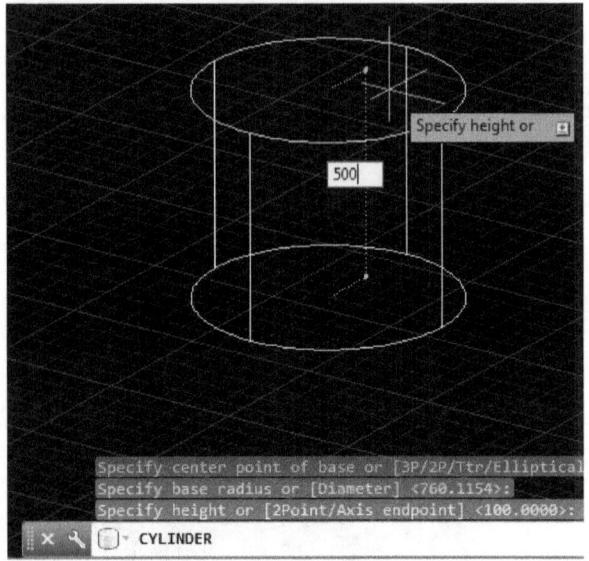

Pic 4.19 Inserting the height of cylinder

5. The result will be like this.

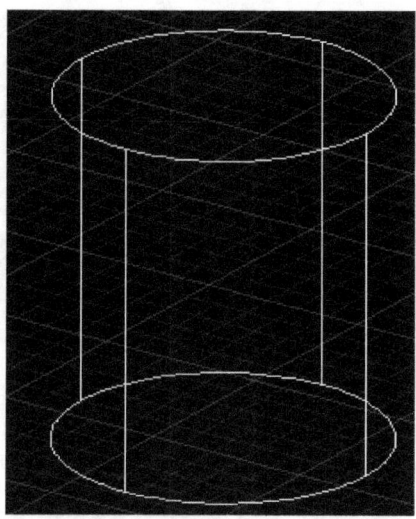

Pic 4.20 Cylinder already created

Another method is by defining radius or diameter. Follow steps below:

1. Repeat steps until step 2

2. Choose d for diameter.

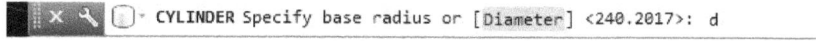

Pic 4.21 Choose D

3. This means, cylinder will be drawn based on diameter.

4. AutoCAD asks for diameter, type: 100.

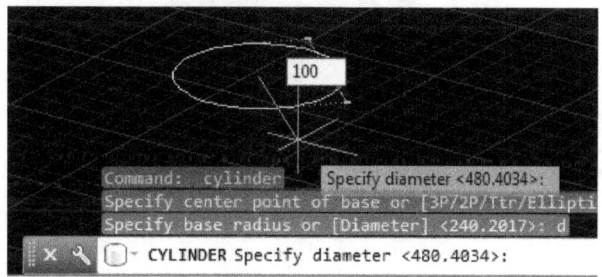

Pic 4.22 Specify the diameter

5. Specify the height: 200.

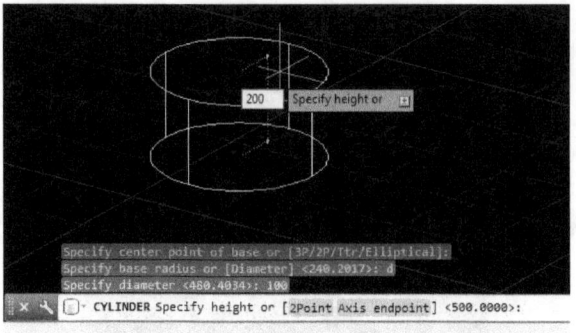

Pic 4.23 Specify the height

6. The result is cylinder with diameter =100 and height = 200

4.2.3 Draw Cone

Cone in AutoCAD can be created using Cone icon. See steps below to create Cone:

1. Click Cone icon.

Pic 4.24 Click on Cone icon

2. AutoCAD asks to specify the center of the circle.
3. Drag mouse to draw a circle.

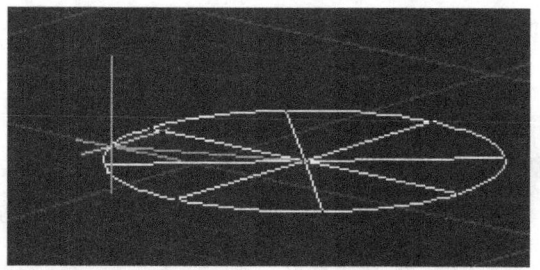

Pic 4.25 Drag mouse to draw a circle

4. AutoCAD asks for height for the cone.

5. The result will be like this.

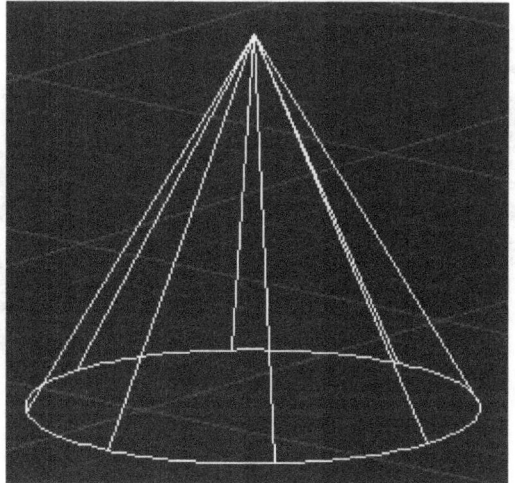

Pic 4.26 Cone result

Or you may specify the diameter and height by using steps below:

1. Repeat steps until step 2

2. In the command prompt, select D to insert diameter.

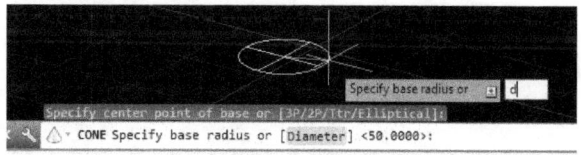

Pic 4.27 Select D

3. It means the circle will be created using diameter.

4. Insert the diameter, for example: 100.

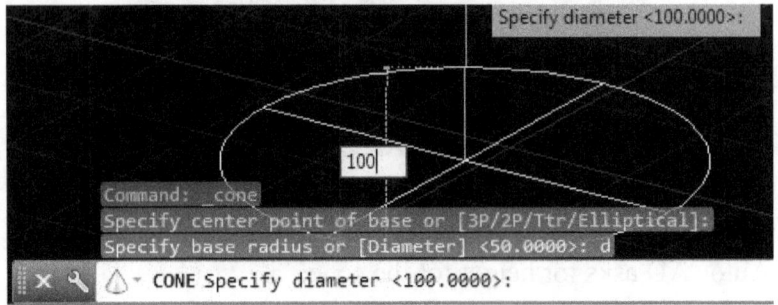

Pic 4.28 Inserting diameter for the circle

5. Specify the height for the cone, for example:150.

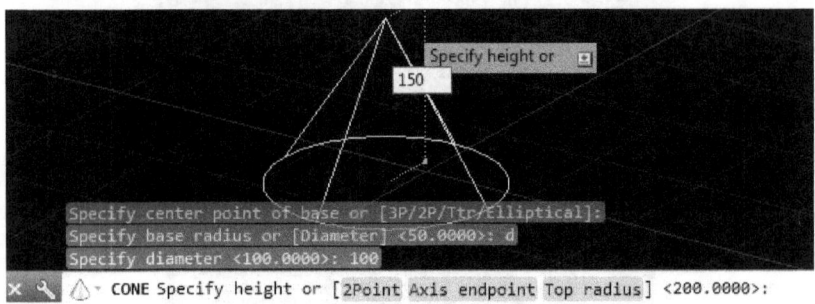

Pic 4.29 Specifying height of the cone

6. The result is cone with diameter = 100 and height = 150

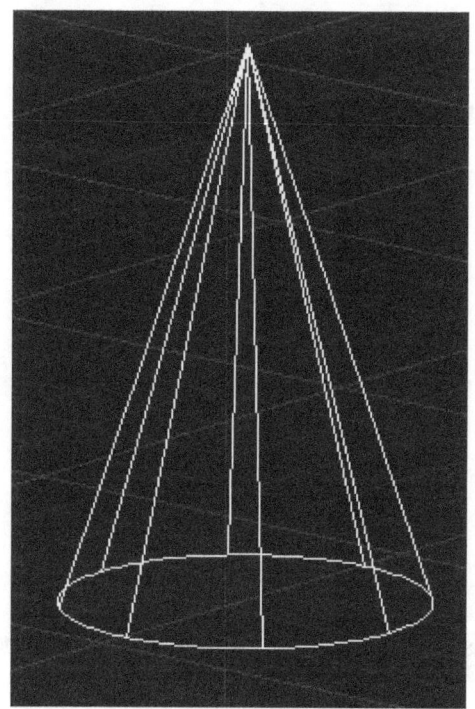

Pic 4.30 Cone result

4.2.4 Draw Ball

Ball can be created using sphere icon. See steps below for drawing ball in AutoCAD:

1. Click Sphere icon.

Pic 4.31 Click Sphere icon

2. AutoCAD asks the circle midpoint. Click the midpoint.

3, Drag mouse to draw the ball.

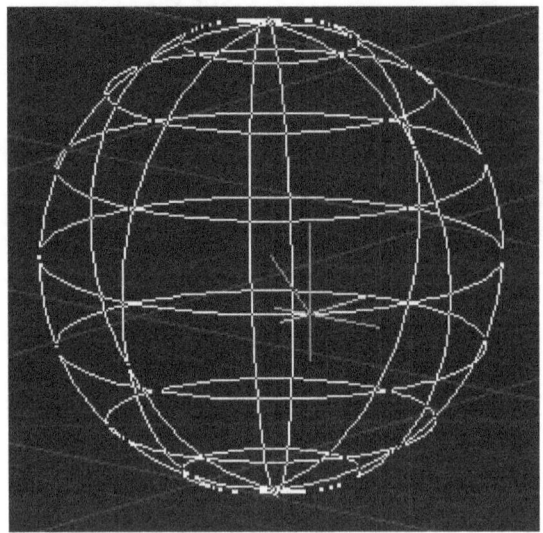

Pic 4.32 Ball result

4. You can see the result on the pictue above.

Another method is by specifying radius or diameter:

1. Repeat steps until step 2.
2. Insert d to specify diameter.

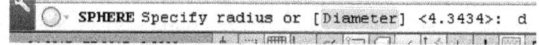

Pic 4.33 Choose Diameter

3. Insert the diameter, for example: 100.
5. The result is a ball in diameter = 100.

4.2.5 Draw Pyramid

You can draw pyramid using steps below:

1. Click pyramid icon,

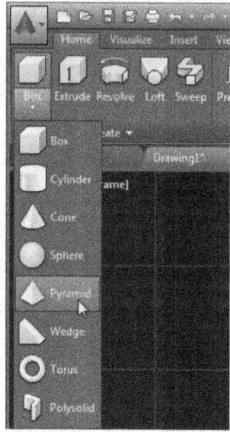

Pic 4.34 Choose Pyramid

2. Specify the center of the rectangle.
3. Drag mouse to create the rectangle.

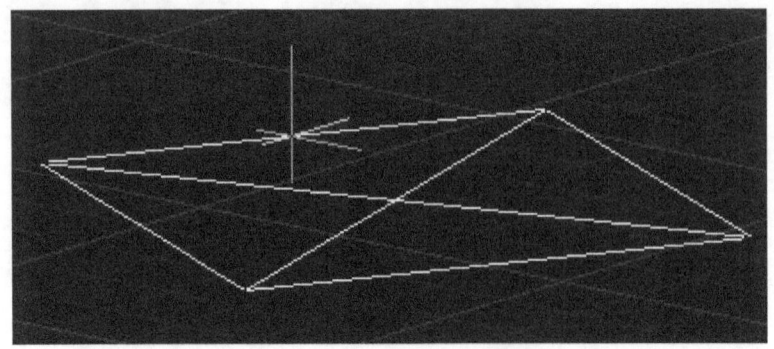

Pic 4.35 Create rectangle

4. Insert the height.

5. See the result in picture below.

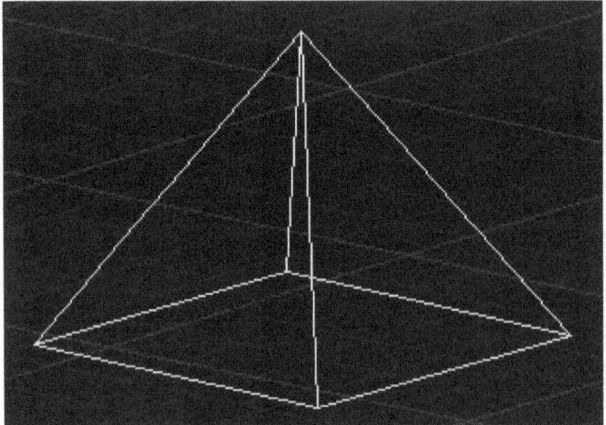

Pic 4.36 Pyramid result

4.2.6 Draw 3D Donut

You can also draw 3D donut using torus command. See example below:

1. Click on **Torus** menu.

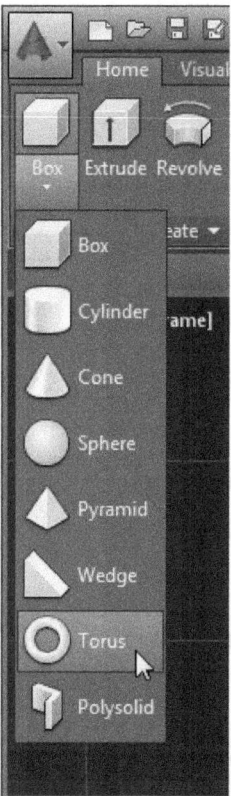

Pic 4.37 Click on Torus menu

2. Enter the center of the circle.

3. Drag the mouse.

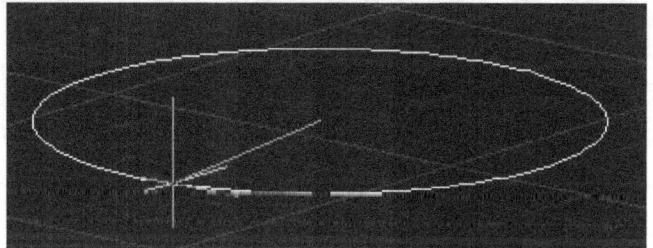

Pic 4.38 Drag the mouse from the center outwards

4. Insert the radius of the small circle of the tube.

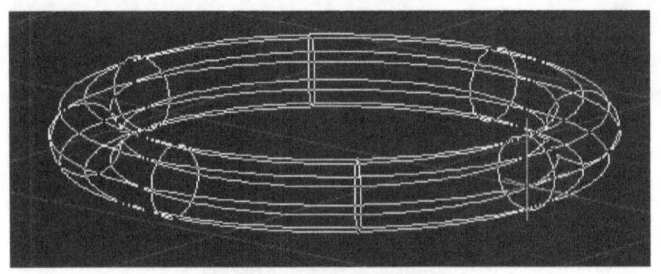

Pic 4.39 Creating the torus

5. Finish.

You can also make torus manually. See steps below:

1. Repeat steps above until step 2.

2. Choose radius, type: 50.

3. The result is a ring in radius = 50.

4. Then specify the radius of the tube = 10.

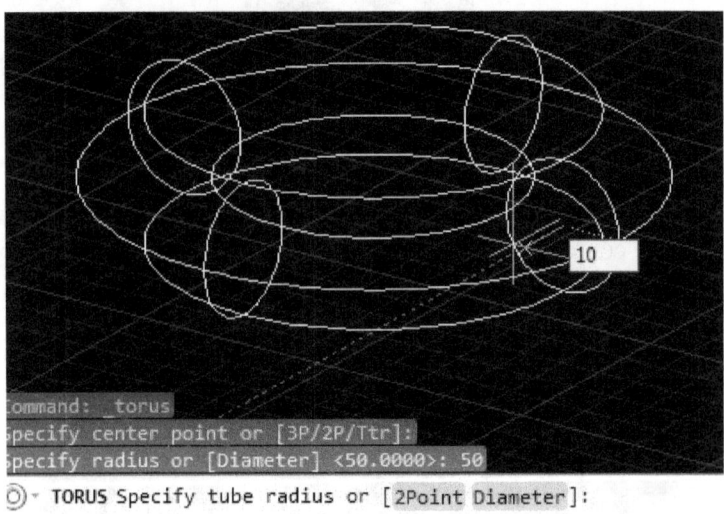

Pic 4.40 Specify the radius of the tube

5. The result is a torus with radius = 50 and tube radius = 10.

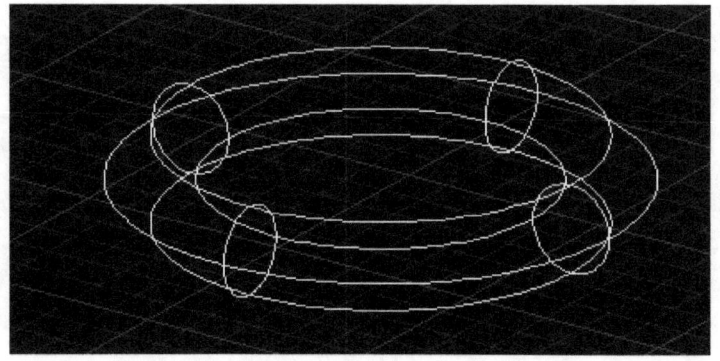

Pic 4.41 Result of ring drawing

4.2.7 Extrude 2D Object

You can draw a 3D object by extruding a 2D object. See example below:

1. Open the 2d pic.

2. Click orbit button.

3. Right click and drag to change the view of the 2d pci.

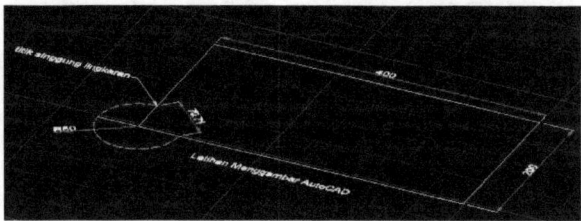

Pic 4.42 2D pic

4. Clik Extrude button like picture below:

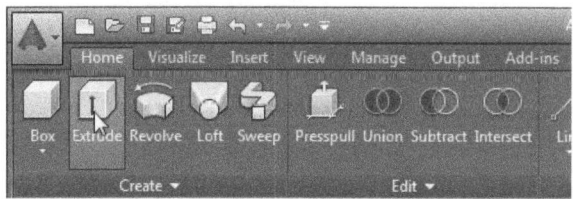

Pic 4.43 Click Extrude button

5. Change the object youw ant to extrude, and click **Enter**.

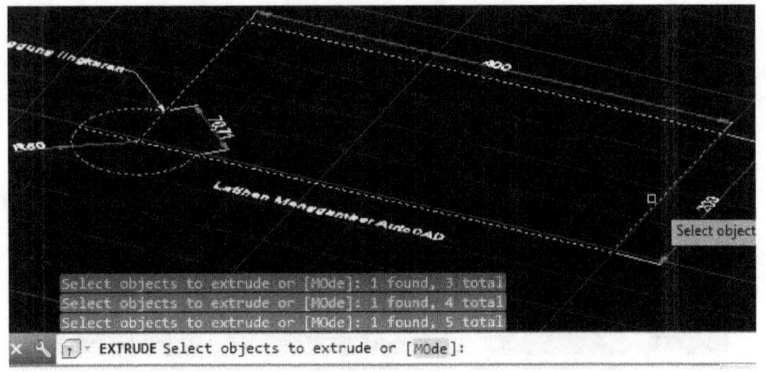

Pic 4.44 Change the object to extrude

6. Insert the height, for example: 1000.

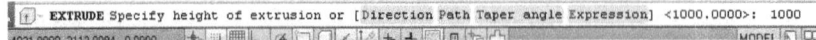

Pic 4.45 Insert the height

7. The 2d pic will be 3D right now, with height = 1000.

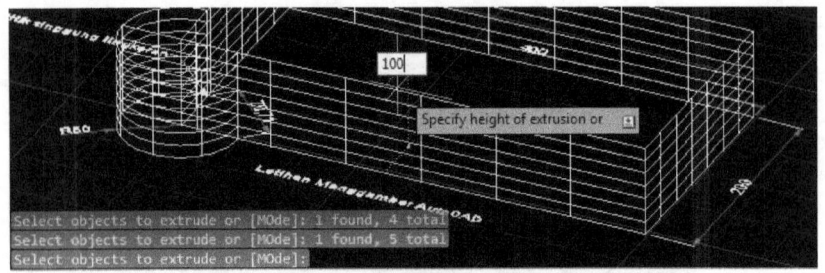

Pic 4.46 Result of extrude process

***

ABOUT THE AUTHOR

Ali Akbar is an AutoCAD Author who has more than 10 years of experience in the architecture and has been using AutoCAD for more than 15 years. He has worked on design projects ranging from department store to transportation systems to the Semarang project. He is the all–time bestselling AutoCAD author and was cited as favorite CAD author. Zico P. Putra is a senior engineering technician, CAD consultant, author, & trainer with 10 years of experience in several design fields. He continues his PhD in Queen Mary University of London. Find out more at https://www.amazon.com/Zico-Pratama-Putra/e/B06XDRTM1G/

CAN I ASK A FAVOUR?

If you enjoyed this book, found it useful or otherwise then I would really appreciate it if you would post a short review on Amazon. I do read all the reviews personally so that I can continually write what people are wanting.

If you would like to leave a review, then please visit the link below:

https://www.amazon.com/dp/B06XS99PKP

Thanks for your support!

www.ingramcontent.com/pod-product-compliance
Lightning Source LLC
Chambersburg PA
CBHW071432180526
45170CB00001B/313